Mathematics Education
for a New Era

Mathematics Education for a New Era

Video Games as a Medium for Learning

Keith Devlin

CRC Press
Taylor & Francis Group
Boca Raton London New York

CRC Press is an imprint of the
Taylor & Francis Group, an **informa** business

AN A K PETERS BOOK

CRC Press
Taylor & Francis Group
6000 Broken Sound Parkway NW, Suite 300
Boca Raton, FL 33487-2742

First issued in hardback 2017

© 2011 by Taylor & Francis Group, LLC
CRC Press is an imprint of Taylor & Francis Group, an Informa business

No claim to original U.S. Government works

ISBN 13: 978-1-138-42788-4 (hbk)
ISBN 13: 978-1-56881-431-5 (pbk)

**Visit the Taylor & Francis Web site at
http://www.taylorandfrancis.com**

**and the CRC Press Web site at
http://www.crcpress.com**

Library of Congress Cataloging-in-Publication Data

Devlin, Keith J.
  Mathematics education for a new era : video games as a medium for learning / Keith Devlin.
    p. cm.
  Includes bibliographical references.
  ISBN 978-1-56881-431-5 (alk. paper)
    1. Mathematics—Study and teaching (Elementary) 2. Mathematical recreations.
    3. Games in mathematics education. 4. Video games and children. I. Title.
QA135.6.D53 2011
372.7–dc22

2010051143

Video games: where art, entertainment, and education can meet

If we are to produce works worthy to be termed "art," we must start to think about what it takes to do so, to set ourselves goals beyond the merely commercial. For we are embarked on a voyage of revolutionary import: the democrative transformation of the arts. Properly addressed, the voyage will lend grandeur to our civilization; improperly, it will create merely another mediocrity of the TV age, another form wholly devoid of intellectual merit.

—Greg Costikyan, game designer with over 30 commercial games to his credit

Contents

 # Preface

This book is written primarily for mathematicians and mathematics educators who want to stay at the head of the development curve in mathematics education. If that is you (and if you are reading this book it probably is), then first and foremost I want to convince you that well-designed video games are going to play a major role in school mathematics education in the future, particularly (though not exclusively) at the middle-school level. My second and broader goal is to initiate a discussion in the mathematics education community about how to rethink mathematics instruction to take advantage of the opportunities for enhanced mathematics learning that video games provide.

Readers do not need any familiarity with video games to appreciate the ideas in this book. Those who are "gamers" (and many mathematicians and mathematics teachers are) can simply skip through the basic descriptions of the medium.

Since video games first began to appear, many educators have expressed the opinion that they offer huge potential for education. The most obvious feature of video games driving this conclusion is the degree to which games engage their players. Any parent who has watched a child spend hours deeply engrossed in a video game, often repeating a particular action many times to perfect it, will at some time have thought, "Gee, I wish my child would put just one tenth of the same time and effort into their math homework." That sentiment was certainly what first got me thinking about educational uses of video games 25 years ago. "Why not make the challenges the player faces in the game mathematical ones?" I wondered at the time. In fact, I did more than wonder; I did something about it, although on a small scale.

This was in the mid 1980s when I was living in the United Kingdom. Soon after small personal computers appeared, so too did video games to play on them. My two (then very young) daughters particularly liked a fast-action game called *Wizard's Lair*, which they played on our first personal computer, an inexpensive machine called the Sinclair Spectrum. *Wizard's Lair* offered a two-dimensional, bird's eye view of a subterranean world. My daughters spent hours playing that game. (A new version of the game, using three-dimensional graphics, was developed and released recently by IGN Entertainment.)

Meanwhile, the brand new computer in their elementary school was sitting largely unused. About the same time that John Sinclair (in due course, Sir John) introduced the Spectrum computer, the British government mandated that every elementary school in the nation be supplied with a computer. (That's right, one *per school*—the early 1980s were the Stone Age of personal computers.) But when my daughters' school received delivery of its shiny new machine, there was no software to run on it. Apparently no one thought of that. Mindful of what I had observed my daughters doing at home, I wrote a simple mathematics education game, in which the player had to use the basic ideas of coordinate geometry in order to discover buried treasure on an island, viewed in two dimensions from directly above. While it did not offer the excitement of *Wizard's Lair*, the children at the school seemed to enjoy playing it. Perhaps offering a prelude of things to come, the game even made a financial profit. I wrote it for free, and the school PTA sold copies to other local schools. I think we sold five or six copies in total, each one on an audiocassette tape. I did not give up my day job.

I thought about writing more games to teach other mathematical topics, but the technology of the day seemed pretty limiting. Perhaps so too was my imagination and innate talent as a game designer, and I did not see a way to present most mathematical concepts and techniques in a way that could not be achieved as well, and in most cases better, by a good (human) teacher. So, after my one brief foray into the game development world, computer games remained for me a parental observation activity until 2003, by which time I was living in the United States and working at Stanford University.

The previous year, Stanford communication professor Byron Reeves and I started a new interdisciplinary research program at Stanford called Media X. Media X carries out research in collaboration with industries, mostly large high tech companies, and one early focus of Media X research was the growing interest in using video-game technology in education. In the fall of 2003, Media X organized a two-day workshop on the Stanford campus called Gaming-2-Learn, to which we invited roughly equal numbers of leading commercial game developers and education specialists from universities interested in developing educational games.

Before I go any further, I should make it clear what I—and the rest of the video game industry—mean by an educational game. As James Paul Gee, among others, points out in his 2003 book *What Video Games Have to Teach Us About Learning and Literacy*, most video games are in fact highly educational. They are all about learning. The question is what is being learned. The term "educational game" is used to refer to games designed explicitly to achieve specific curricular goals in school subjects like math and science. We'll meet Gee again several times throughout this book.

Getting back to the Gaming-2-Learn workshop, to kick off the conference I conducted an on-stage discussion with leading game developer Will Wright, founder of the game company Maxis (now owned by Electronic Arts) and developer of the bestselling simulation (or "life world") games *SimCity*, *The Sims*, and *Sims Online*, and more recently the evolution-inspired MMO *Spore*. The message for the education community from Wright and the other game developers at the conference was not encouraging. They indicated that developing a good game was still largely a hit-and-miss affair that required a huge investment of time and money (with budgets often in excess of $10 million for a single game). Meeting educational goals while producing a compelling game that would sell sufficiently well to recoup the huge investment costs in such a highly competitive and risky market was generally viewed as a nonviable commercial proposition. The professional game developers were reluctant to divert their attention from an increasingly profitable new branch of the entertainment industry to go out on a limb and produce an educational game.

That is not to say that highly sophisticated, professionally produced educational games have not been developed. In particular, the US Army has spent many tens of millions of dollars to develop immersive three-dimensional games to recruit and to train soldiers. Apart from a few special cases, however, the development of educational games has been largely left to the educators themselves, generally working with a small grant rather than a large commercial development budget. It is no criticism of the pioneering educators who have taken those first steps that the results have not been in the same league as the entertainment games produced by game development professionals.

Unfortunately, with commercial entertainment games getting better all the time, the bar for an educational game to be successful is being continually raised, as the expectations of the target audience (students) go up with every new game innovation.

Following the Gaming-2-Learn event, I started to take a closer look at the educational video games that were starting to appear. Most focused on basic skills, and that remains true today, but as I'll explain in the next chapter, mathematical thinking is much more than mastery of basic skills. What is needed, if we are to take full advantage of the educational potential video games offer, are games that develop conceptual understanding and genuine mathematical thinking. This book, conceived back then but only now coming to fruition, is intended to help us move toward that goal.

Something else I started to do after Gaming-2-Learn was learn something about video game design. There weren't many books back then but there were a lot of websites and blogs that talked about the process. Taking advantage of my location in the heart of Silicon Valley, I also had many conversations with professional video game designers, giving me insights into the organizational and engineering complexities involved in commercial video game production.

Overall, I learned a lot in that period, but the message from the Gaming-2-Learn conference remained pertinent: it took a lot of money and effort to design and build a good video game, and no one knew for sure it would be successful until it was finished and released. That perhaps explained why the math ed games that were coming out looked like forced marriages of video games with traditional instruction of basic skills; it was hard enough just getting a game out, without trying to integrate mathematical thinking into the game play.

But with the cost of developing games steadily coming down, we are rapidly approaching a time when individuals and small groups with fairly limited resources can produce good mathematics education video games that develop more than basic skills. We can expect better and better mathematics education video games to appear over the coming years, and as they do, so will they will play an increasing role in mathematics education. This book is written for teachers who will soon need to know how to take advantage of mathematics education video games, for curriculum developers who will soon need to take account of video games, for teachers and others who may be tempted to develop effective mathematics educational games, and for anyone who is curious to know how mathematics education is going to look a decade from now and wants to understand why.

I should perhaps mention one more thing that I did after the Gaming-2-Learn workshop: I became a gamer. In 2005, I started to play *World of Warcraft*, a game that at the time had garnered just over a million players worldwide, and was on its way to becoming a global sensation. Today it has over 15 million players and has been the subject of many studies. *World of Warcraft* is a massively-multiplayer online role-playing video game, or MMO. MMOs are typically played on a personal computer. I'll say more about the different genres of video games later in the book. *World of Warcraft* remains my favorite game today, and at the time of writing, I am a master level player.

Conversations with a great many people over the years have contributed to this book, but I want to express particular appreciation to Johnny Lott, professor emeritus of mathematics education at the University of Mississippi and a former president of the National Council of Teachers of Mathematics, Professor John Mason of the Open University in the UK, and Professor Anne Watson of Oxford University in the UK, each of whom read through the entire manuscript and offered many helpful suggestions.

Finally, thanks also to Klaus Peters, my publisher. Klaus published my first book in 1973, and we have been good friends ever since. It is always a pleasure to work with him and his team.

—Keith Devlin
Palo Alto, CA
January 2011

 # 0 + 1 State of Play

The United States ranks much lower than most of our major economic competitors in the mathematics performance of high-school students. That national slide begins in middle school. Many attempts have been made to improve US middle-school mathematics education, but most have failed to achieve the desired results. I think the reason is clear. They have focused on improving basic math skills. In contrast, I, and a great many of my colleagues, believe the emphasis should be elsewhere. Mathematics is a way of thinking about problems and issues in the world. Get the thinking right *and the skills come largely for free*. This simple phrase, while correct as far as it goes, requires some elaboration.

First, what exactly is a basic skill? As a rough rule of thumb, the basic skills are the computations that can become so routine that they can be performed in an almost subconscious fashion. For example, most people would say the basic skills include knowledge of the multiplication tables and an ability to perform addition, subtraction, and multiplication of whole numbers and fractions. But what else? Handling parentheses in manipulating arithmetic or algebraic expressions is another skill that I think most educators would say is important. What about long division? In the pre-calculator era it was a very important skill to master. But today many mathematics educators would probably say this is not an essential basic skill, though the process of acquiring mastery does lead to an important understanding of place-value arithmetic and experience with a computational algorithm.

Second, my claim that the basic skills follow "largely for free" from understanding does not mean a student does not have to practice those skills. I'll come to this issue in due course. The complexity of the relationship between the development of mathematical thinking and the mastery of basic skills becomes clear when you recognize that the former requires the latter.

There are two reasons why the focus of mathematics education video games has until now been largely on skills. First, many people—even those in positions of power and influence—don't understand what mathematics is and how it works. All they see are the skills, and they think—wrongly—that mathematics is about those skills. Given that most people's last close encounter with mathematics was a skills-based school math class, it is not hard to see how this misconception arises. But to confuse mathematical thinking with mastery of basic skills is akin to confusing architecture with bricklaying, or playing a musical instrument with being able to play the musical scale. Of course you need the basic skills. You can't build a brick house unless you know how to lay bricks, and you can't play an instrument if you don't know how to sound the different notes. Similarly, you can't think mathematically if you have not mastered the basic skills. But mathematical thinking is far more than merely having the basic skills at your fingertips, just as architecture is more than laying bricks and music more than playing notes.

This comparison should not be taken as a suggestion that students should first master basic skills and then develop mathematical thinking. The former is indeed required for the latter, but the skills are much more easily acquired when encountered as a part of mathematical thinking. In my opinion, viewing these issues simplistically and separately is a large part of what is wrong with much mathematics instruction in the United States. The two are each major components of a symbiotic whole. The same is true for learning to play a musical instrument. Doing so requires an ability to play the right notes, but that basic musical skill is acquired far more easily and effectively by trying to play tunes than by endlessly practicing the scales.

The other reason for the continued focus on skills is more substantial. For over 2000 years, the only way to provide mathematics education to everyone was through the written word—textbooks. But in order to learn mathematical thinking from a textbook, you have to approach it through learning the skills. That means you have to master the skills first. And for reasons I just indicated, that is not an effective educational approach.

So mathematics is not about acquiring basic skills or learning formulas. It's a way of thinking about problems in the world. The skills are merely the tools you need in order to do that thinking. Math is not a body of knowledge, it's something you *do*. And the printed word can be a terribly inefficient way to learn how to do something. The best way for an individual to learn how to do something is, as the Nike slogan says, "Just do it!" For example, if you want to learn to play chess, you can learn the rules from a book, but you won't learn to play chess until you start playing games. The same is true for learning to ride a bicycle, learning to swim, to ski, to play tennis, or to play the guitar.

To be sure, for some activities it is hard—or even dangerous—to learn by doing. Airlines train their pilots to fly aircraft not by putting them in control of a real airplane but by giving them time in a flight simulator. NASA trains its astronauts the same way. Medical schools now train surgeons using simulation. And the US Army uses simulators to train soldiers before they go anywhere near the battlefield. This is the way we should be teaching middle-school mathematics. Sound crazy? If it does, it's because you probably have a false impression both of the technology I have in mind and of what mathematics is. In this book, I'll try to correct both false impressions.

Of course, the best way to demonstrate what I have in mind is to build one or more mathematics simulators so you can experience the technology. But even without doing that, there is a very good reason to expect the approach I am advocating to be extremely effective. Why? Because today's students have a natural affinity for this technology. There is another name for digital simulators. We call them video games. True, for some readers—probably many over the age of 40—that term might evoke an image of fast-action shooter games. But remember, books range from trashy novels to great works of art and, perhaps more to my point, to textbooks.

The kind of learning that is possible in a simulator—namely, through guided exploration and experience—has long been known to be extremely effective. The fancy name for this kind of process is "situated learning." It works because it is something that evolution has equipped humans to do. We do it naturally and spontaneously. For example, numerous studies over many years have shown that ordinary people of all ages typically achieve mastery of middle-school mathematics when they need it to do something that matters to them in the course of their everyday lives. They master it rapidly to a level of 98 percent accuracy (see Chapter 2). So why haven't we been using situated learning to teach middle-school math for generations? Because in the physical world it's impossible to provide such an education except on a one-on-one basis in some form of apprenticeship or tutoring system. With video game technology, however, you can do it.

Math Ed Video Games: The Current State of Play

At this stage, I suspect I need to take a moment to respond to a question that may have been on your mind since you first opened this book. It comes up a lot. Whenever I tell someone that I am studying mathematics education (math ed) video games, if they say anything at all, it is to ask me if I have seen or played game X and what do I think of it. Often I have, and I have to tell them that game X isn't really teaching or helping students learn mathematical thinking; rather, it is focused on trying to develop mastery of basic skills. It usually doesn't matter which game X is mentioned. With just a few, little-known exceptions, that's what they all do! Even the ones that claim they develop "deep conceptual thinking" are, in their implementation, for the most part doing little more than providing an opportunity to practice basic skills.

Now, there is nothing wrong with practicing basic skills. Without mastery of basic skills, you cannot progress very far in mathematics. But mastering skills is much more rapidly achieved, and with longer lasting effect, if that practice is done in the pursuit of a meaningful goal rather than mere repetition for its own sake, and a video game can provide that element. But as I stressed earlier, basic skills are just the raw materials out of which useful mathematical thinking can be developed. The main reason that most of the current games concentrate on skills practice is that games to practice skills are far easier to construct than games that develop genuine mathematical thinking.

Just as it is a lot easier to teach and learn bricklaying than architecture, or to achieve mastery of the musical scale than writing or playing a sonata, so too it is a lot easier to teach and learn basic skills than to teach and learn mathematical thinking. As a result, when a new technology comes along with educational potential, like video games, it's not surprising that entrepreneurs first go after the low hanging fruit. Thus the rash of video games that help students master basic skills.

I stress that to date I have not seen a single video game that really develops mathematical thinking. A few do develop logical thinking, such as the puzzle game *XOR* created by Astral Software in 1987 for a range of platforms. But those games never developed mass appeal. Having spent many years studying video games and talking with video game designers, I fully understand why this is the case. It is

Figure 1. The ornithopter: an early attempt to fly.

Figure 2. *Timez Attack* and *DimensionM*, two early attempts at math ed video games.[1]

much harder to provide an environment that develops mathematical thinking than an environment in which students can merely practice their basic skills.

When I see or play most current math ed video games, I am reminded of the image shown in Figure 1.

When people made the first attempts to fly, the most successful machines for transport were wheeled vehicles, and the only known examples of flying creatures were birds and insects, both of which fly by flapping wings. Taking those as their starting point, early aspiring aviators took bicycle-type machines and added bird-like wings that were flapped by some sort of body motion. The results were uniformly disastrous. Wings may be how birds and insects fly, but that doesn't work for humans. The key to human flight was to separate flying from flapping wings, and to achieve flight by another means more suited to machines built from wood or metal.

Figure 2, which shows two examples of recent attempts to build math ed video games, illustrates my point. First, I should say that these are two of the best math ed video games on the market today, and both have achieved respectable sales. The one on the left, *Timez Attack*, has the modest goal of providing kids with an engaging way to practice their multiplication tables; the player has to key in the correct answer to a series of multiplication problems in order to kill the monster before it kills the player. The one on the right, *DimensionM*, focuses on algebra and the rules of arithmetic. The player has to answer the questions correctly to progress in the game. Both games probably achieve the goals their developers set for their products. (It would require research to determine the efficacy and indirect consequences of any new form of instruction. For example, does a monster-killing format overly encourage fast action and rapid recall at the expense of reflection as a general strategy?)

[1] *Timez Attack* is by the company Big Brainz and *DimensionM* by Tabula Digita.

I think both examples are reasonable early-generation math ed video games. But the screen images in Figure 2 remind me of the ornithopter in Figure 1. (Though to be fair, unlike the ornithopters those video games do manage to get off the ground.) Let's face it, in real life you are rather unlikely to find yourself face to face with a person or a monster who has a math problem written on their chest, or a situation in which a math formula suddenly appears before your eyes, hovering in the air. Symbolic expressions are the way people do math when they are working with paper and pencil. They look out of place in a game world. And the reason they look out of place in that environment is because they *are* out of place.

Video game worlds are not paper-and-pencil symbolic representations; they are imaginary *worlds*. They are meant to be lived in and experienced. (Not all video games have worlds. But any game is played in an environment, in part created by the game, that affects the action. If you don't believe this, pay particular attention next time you see someone playing a casual game like *Bejeweled* on a smartphone. I'll discuss the special case of those kinds of games in the final chapter.)

Putting symbolic expressions in a math ed game environment is to confuse mathematical thinking with its static, symbolic representation on a sheet of paper, just as the early aviators confused flying with the one particular representation of flying which they had observed. To build truly successful math ed video games, we have to separate the activity—a form of thinking—from its familiar representation in terms of symbolic expressions.

Mathematical symbols were introduced to do mathematics first in the sand, then on parchment and slate, and still later on paper and blackboards. *Video games provide an entirely different representational medium.* As a dynamic medium, video games are far better suited in many ways to representing and doing middle-school mathematics than are symbolic expressions on a page. We need to get beyond thinking of video games as an environment that delivers traditional pedagogy—a new canvas on which to pour symbols—and see them as an entirely new medium to represent mathematics. That is in my view the single most important message of this book.

Now that I have provided you with a rough road map of where we are heading, I'll explain to you why video games are the way to go. The story begins twenty years ago in the streets of South America.

 # Street Smarts

The young woman walks up to a stall piled high with coconuts. Behind it stands a young boy of around 12 years of age, who is taking care of the stall while his parents have stepped away for a moment. It's hot and there is a lot of noisy activity in the market, one of several in the Brazilian city of Recife.

"How much is one coconut?" the woman asks.

"Thirty-five," the boy replies with a smile.

"I'd like ten. How much is that?"

The boy pauses for a moment before replying. Thinking out loud, he says: "Three will be 105; with three more, that will be 210. (Pause) I need four more. That is ... (pause) 315 ... I think it is 350."

Though the boy gets the answer right, the woman can't help but wonder why he did not use the simple rule that to multiply by 10 you just add a 0, so ten coconuts at 35 Cruzeiros each will cost Cr$350.

A short while later, at another stall, again staffed by a young boy, this one about 14 years old, the woman makes a purchase that requires the child to subtract Cr$75 from Cr$243. The boy calculates out loud:

"You just give me the two hundred. I'll give you twenty-five back. Plus the forty-three that you have, the hundred and forty-three, that's one hundred and sixty-eight."

If you were the shopper, faced with a young child at a stall in a noisy, busy South American street market who calculated your change in that fashion, you might suspect that the young salesman was trying to pull a fast one. In fact, his answer is perfectly correct. In a moment I'll examine what he is doing.

7

Incidentally, you may think it strange that a book about using video games in education begins the pedagogy section with a discussion about street markets. You may even be tempted to skip the rest of this chapter and look for the "new stuff" about video games. However, what you will find in this chapter is crucial for the entire remainder of the book, and sets the stage for everything else.

Returning to those two marketplace exchanges, I am not making them up. They are taken verbatim from a report written some years ago by three researchers, Terezinha Nunes (the "young woman" in my story), then at the University of London, England, now at Oxford University), and Analucia Dias Schliemann and David William Carraher of the Federal University of Pernambuco in Recife, Brazil.[1] The three researchers went out into the street markets of Recife with concealed tape recorders, posing as ordinary market shoppers. They sought out stalls being staffed by young children between 8 and 14 years of age. At each stall, they presented the young stallholder with a transaction designed to test a particular arithmetical skill. The purpose of the research was to compare traditional instruction (which all the young market traders had received in school since the age of six) with learned practices in context.[2]

How well did our two young sellers do? Let's start with the coconut seller, who did not appear to know the rule that to multiply by 10 you simply add a 0—so 35 becomes 350. It turns out that this simple trick isn't the only mathematical fact he did not know. Despite spending six years in school, he had almost no mathematical knowledge at all in the traditional sense. When given school-type tasks in a school-type setting he performed badly. What arithmetical skills he had were self-taught at his market stall. Here is how he solved the problem.

Because he often found himself selling coconuts in groups of two or three, he needed to be able to compute the cost of two or three coconuts; that is, he needed to know the values $2 \times 35 = 70$ and $3 \times 35 = 105$. Faced with a highly unusual request for ten coconuts—how was the woman going to carry ten coconuts and what was she going to do with them, by the way?—the young boy proceeded like this. First, he split the 10 into groups he could handle, namely $3 + 3 + 3 + 1$. Arithmetically, he was now faced with the determining the sum $105 + 105 + 105 + 35$. He did this in stages. With a little effort, he first calculated $105 + 105 = 210$. Then he computed $210 + 105 = 315$. Finally, he worked out $315 + 35 = 350$. Altogether quite an impressive performance for a twelve-year-old who supposedly "couldn't do math"!

[1] T. Nunes, A. Schliemann, and D. Carraher, *Street Mathematics and School Mathematics*, Cambridge University Press, Cambridge, UK, 1993.
[2] A similar study I'll mention later was described in J. Lave, *Cognition in Practice: Mind, Mathematics and Culture in Everyday Life*, Cambridge University Press, Cambridge, UK, 1988. More generally, see J. S. Brown, A. Collins, and P. Duguid, "Situated Cognition and the Culture of Learning," Educational Researcher 18 (1), 1989, pp. 32–41.

What about the second boy, the one with the whiz-bang answer to 243 – 75? First, it's clear from what he went on to say that his first sentence was meant to be "You just give me the one hundred." For what he was doing was splitting up the 243 into 100 + 100 + 43. He put the 43 and one of the 100s to one side and subtracted the 75 from the remaining 100. That's something he could do easily: 100 – 75 = 25. (Presumably it's a result he had seen so often that he knew it by heart.) Then he added back the 43 and the 100. To do this, he first computed 100 + 43 = 143 and then calculated 25 + 143 = 168. That last step was still a challenging addition, of course, and the boy struggled with it. But in the end he got the right answer.

In essence, what his overall method did was change the challenging subtraction problem 243 – 75 into the addition problem 143 + 25, by subtracting 100 from 243 and adding 100 to 75. The final addition was not an easy one, but his method worked because, like most people, he found addition much easier than subtraction.

Pretty remarkable, don't you think? But there's more. Posing as customers was just the first stage of the study Nunes and her colleagues carried out. About a week after they had "tested" the children at their stalls, they went back to the subjects and asked each of them to take a pencil-and-paper test that included exactly the same arithmetic problems that had been presented to them in the context of purchases the week before.

The investigators were careful to give this second test in as non-threatening a way as possible. It was administered in a one-on-one setting, either at the original location or in the subject's home, and included both straightforward arithmetic questions presented in written form and verbally presented word problems in the form of sales transactions of the same kind the children carried out at their stalls. The subjects were provided with paper and pencil, and were asked to write their answer and whatever working they wished to put down. They were also asked to speak their reasoning aloud as they went along.

Although the children's arithmetic had been close to flawless when they were at their market stalls—just over 98% correct—they averaged only 74% when presented with market stall word problems requiring the same arithmetic, and a staggeringly low 37% when the same problems were presented to them in the form of a straightforward symbolic arithmetic test.

The performance of our young coconut seller was typical. One of the questions he had been asked at his market stall, when he was selling coconuts costing Cr$35 each, was: "I'm going to take four coconuts. How much is that?" The boy replied: "There will be one hundred five, plus thirty, that's one thirty-five . . . one coconut is thirty-five . . . that is . . . one forty." Let's take a look at this solution. Just as he had in the transaction I described first, the boy began by breaking the problem up into simpler ones; in this case, three coconuts plus one coconut. This enabled him to start out with the fact he knew, namely that three coconuts cost Cr$105. Then, to add on the cost of the fourth coconut, he first rounded the cost of a coconut to Cr$30 and added that amount to give Cr$135. While not verbalizing the next step precisely, he noted that the "correction factor" for the rounding was Cr$5, and added in that correction factor to give the correct answer of Cr$140.

On the formal arithmetic test, the boy was asked to calculate 35×4. He worked mentally, vocalizing each step as the researcher had requested, but the only thing he wrote down was the answer. Here is what he said: "Four times five is twenty, carry the two; two plus three is five, times four is twenty." He then wrote down 200 as his answer. Despite the fact that, numerically, it was the same problem he had answered correctly at his market stall, he got it wrong. If you follow what he said, it's clear what he was doing and why he went wrong. In trying to carry out the standard right-to-left school method for multiplication, he added the carry from the units-column multiplication (5×4) *before* performing the tens-column multiplication rather than afterwards, which is the correct way. He did, however, keep track of the positions the various digits should occupy, writing the (correct) 0 from the first multiplication after the (incorrect) 20 from the second, to give his answer 200.

The same thing happened with another child stallholder, this time a girl of nine. When a researcher approached the child at her coconut stall and asked, "I'll take three coconuts. How much is that?" the young seller replied, "Forty, eighty, one twenty." With one coconut costing Cr$40, her technique was to keep adding 40 until she reached the correct number of additions. On the school-like arithmetic test, the same girl was presented with the multiplication 40×3. Her answer was 70. Her explanation of how she arrives at that answer was: "Lower the zero; four and three is seven."

Despite the fact that she had no trouble operating a stall in a noisy, busy street market, the young girl's recollections of the standard arithmetical procedures she had been taught in school were mired in confusion. How bad was her confusion? The same girl, upon being asked for 12 lemons priced at Cr$5 each, separated them out two at a time, saying as she did so, "Ten, twenty, thirty, forty, fifty, sixty." But when she was presented with the problem 12×5 on the test—numerically the very same computation—she first lowered the 2, then the 5, and then the 1, giving the answer 152.

A similar degree of confusion about school arithmetic was exhibited by another child street-seller who had no trouble with a subtraction task when it had arisen at the market stall, but went badly awry when presented with the equivalent addition on the school-like written test. At the market stall, where the boy had been selling coconuts for Cr$40 each, the customer paid with a Cr$500 bill, and said, "I'll take two coconuts. What do I get back?" "Eighty, ninety, one hundred, four twenty," the boy replied.

On the test, the child was presented with the addition 420 + 80. He gave the answer 130, apparently proceeding as follows: add 8 to 2 to give 10; carry the 1; add 1 (the carry), 4, and 8, to give 13; write down the final 0 in the units column to give 130. Eventually, with some prodding by the researcher, the boy was able to reach the right answer—by ignoring the pencil and paper and using a counting on method.

A similar outcome arose in another case, after a subject had failed to solve the division problem: 100/4. She first tried to divide 1 by 4, then tried to divide 0 by 4, and then gave up, claiming that it was not possible. Prodded by the researcher, she replied: "See, in my head I can do it . . . Divide by two, that's fifty. Then divide by two, that's twenty-five."In other words, she used the fact that dividing by 4 can be achieved by dividing by 2 twice in succession—together with her ability to halve the numbers 100 and 50.

In case after case, Nunes and her colleagues obtained the same results. The children were absolute number wizards when they were at their market stalls, but virtual dunces when presented with the same arithmetic problems presented in a typical school format. The researchers were so impressed—and intrigued—by the children's market stall performances that they gave it a special name: they called it *street mathematics*.

Street Mathematics

Street mathematics is the mathematics that people develop for themselves, when they need it. It is not restricted to young market traders in Brazil, and you can find it in other locations besides the streets. For instance, you can find it in the United States, as schoolteacher James Herndon described in his 1971 book *How to Survive in Your Native Land.*[3]

Herndon recounts how, on one occasion, he was teaching a junior high school class of children who had all essentially failed in the school system. At one point, he discovered that one of the students had a well-paid, regular job scoring for a local bowling league, a task that required fast, accurate, and complicated arithmetic. (Have you ever seen the scoring system in bowling?) Seeing a golden opportunity to motivate this student to do well in class, Herndon created a set of "bowling score problems" and gave them to the boy. The attempt was a complete failure. During evenings in the bowling alley, the boy could keep accurate track of eight different bowling scores at once. But he could not answer the simplest scoring question when it was presented to him in the classroom. In Herndon's own words, "The brilliant league scorer couldn't decide whether two strikes and a third frame of eight amounted to eighteen or twenty-eight or whether it was one hundred eight and a half."

---

[3] Simon and Schuster, New York, 1971.

Herndon observed similar failure when he tried to reach other students in the class by presenting them with problems of the very kind they solved with ease outside the classroom. For example, to a girl who admitted she never had any trouble shopping for clothes, he gave the problem: "If you buy a pair of shoes costing $10.95, how much change do you get from a twenty?" The girl answered "$400.15," and wanted Herndon to tell her if it was right!

Since both the Recife children and Herndon's students demonstrated that they could handle arithmetic in the appropriate context, when the numbers meant something to them, and when the consequences mattered to them, it seems clear that meaning and motivation play major roles in our ability to do arithmetic. (These may not be the only factors at work. Social and cultural circumstances may also play an important attitudinal role, and I'll discuss these and other issues affecting performance in due course.)

Achievement level was not the only difference between street mathematics and school mathematics. The transcriptions of the verbal market stall exchanges also showed that the children's street methods of computation were different from those taught at school. Yet the school methods are taught in part because they are supposed to be easier! Indeed, for anyone who masters both methods, those taught in school *are* easier—just compare the method our first subject used to compute 10×35 with the schoolroom method for solving the same problem. Moreover, the school methods are much more general than some of the "quick tricks" picked up on the job, and therefore can be applied much more widely. Seeking powerful, efficient, and more general methods is an important part of mathematics.

Why did the people who used street mathematics seem to ignore the standard methods? Intrigued by this question, Nunes and her colleagues set out to examine the methods used by the child stallholders. The researchers' approach was to determine the difference between the children's abilities in mental (or oral) arithmetic and written arithmetic, when both were measured under test conditions. The children never performed as well during formal testing as they did when at work at their stalls. Nunes and her colleagues wanted to know if there was a measurable difference between the two ways of doing arithmetic on a test, and sought to explore how the *methods* of street mathematics and school arithmetic differed.

The group of children that Nunes and her colleagues tested consisted of 16 students, including boys and girls. All were in the third grade at school, where they had been taught the standard procedures for addition, subtraction, multiplication, and division. Because many children in Brazil have to repeat the same grade level two or more times, the ages of the children ranged from 9 to 15. The older children had not only more years instruction at school arithmetic, they had also spent longer working in the street market. The subjects were given three kinds of problems: simulated sales transactions of the kind they were familiar with in the market, word (or story) problems, and straightforward computational arithmetic problems. In all but one category, the children performed better at mental arithmetic than they did with pencil and paper. In most cases, the differences were dramatic.

In the case of addition, for the simulated sales questions, the children averaged 67% correct orally and 75% correct on the written test. This was the only case where their pencil-and-paper results were better than their oral answers (i.e., the answers they obtained by working in their heads without the aid of pencil and paper). For the addition word problems they averaged 83% correct orally and just 62% written. For the straightforward computation questions, they got a perfect 100% orally compared with a significantly lower 79% in writing.

For subtraction, the difference between their oral performance and their written performance was striking for all three kinds of questions. In the simulated sales they averaged a so-so 57% correct orally (far less than when calculating change at their stalls) and a mere 22% correct in writing. For the word problems, the figures were a moderately good 69% orally and that same low 22% in writing. For the computation problems, their performances were 60% correct orally—not too bad—but a miserable 14% in writing.

For multiplication, the corresponding figures were a comfortable 89% correct orally and a disappointing 50% correct in writing for the simulated sales, 64% orally and 50% in writing for the word problems, and a perfect 100% orally against a poor 39% in writing for the computation problems.

When it came to division, the results were extremely poor. The children averaged 50% correct orally on all three kinds of problems, but they had clearly failed to master the schoolroom method of division. When asked to answer the questions using pencil and paper, they scored 0% correct on the simulated sales and the word problems, and got just 7% correct on the straightforward division questions. In short, the children could not do division under any sort of test conditions.

Clearly, the children were much better at mental arithmetic than they were at applying the paper-and-pencil methods they had been taught at school. (Presumably the same will be true of anyone who makes regular use of numbers and basic arithmetic in their lives.) But there is still the question of how they were achieving their much greater success in oral arithmetic compared to written arithmetic. Since they appeared unable to use the methods they had been taught in school, just *how* were these children solving the problems when they worked them in their heads?

You get some idea of the children's methods—and hence a first indication that street mathematics is something very different from school arithmetic—when you look at the transcripts of what the children actually said as they were working out the problems mentally. Their words reveal that they were using some sophisticated manipulations of numbers. For example, when faced with computing 200 – 35, one child proceeded like this: "If it were thirty, then the result would be seventy. But it's thirty-five. So it's sixty-five. One hundred sixty-five."

Here is what he was doing. First he split the 200 into 100 + 100. (He did not vocalize this step, but it's clear from what came after that this is what he did.) He put one 100 to one side and set out to compute 100 – 35. To do this, he first rounded off 35 to 30, and computed 100 – 30. This he could do easily; the answer is 70. Then he corrected for the rounding by subtracting the 5 he had ignored; 70 – 5 = 65. Finally, he added the 100 he had put to one side at the beginning: 65 + 100 = 165.

Here is one more example, this time involving division. As we saw earlier, most of the children had significant difficulty with division when working orally and failed completely when trying to use the school-taught procedure. The problem was to calculate 75/5, asked as a question about sharing 75 marbles among 5 children. One child said, "If you give ten marbles to each, that's fifty. There are twenty-five left over. To distribute to five boys, twenty-five, that's hard. . . . That's five more for each. Fifteen each." Absolutely right! The child began by "rounding" 75 to 50 and solving the simpler problem 50/5, for which he had no trouble computing the answer 10. He appeared to know that as a fact, which is why he performed the initial rounding down from 75 to 50. The rounding left 25 marbles still to distribute. He found this difficult; he did not know the answer to 25/5. But after a bit of thought he figured it out: 25/5 = 5. Now all he had to do was add that 5 to his previous result of 10 to give his final answer, 15.

Grown-Ups Too

It's not just children that exhibit a huge disparity between the math they are able to do in the everyday world and their poor performance when presented with a "math test." I'll describe one particular study of adults because it illustrates well an additional finding of such research that I believe will be important in the design of any video game intended to develop mathematical thinking.

In the early 1980s, the anthropologist Jean Lave carried out a study called the Adult Math Project (AMP).[4] Currently a faculty member in the Department of Education at the University of California at Berkeley, Lave was at the University of California at Irvine at the time of this study. The subjects she studied were ordinary people in Southern California, shopping in a supermarket.

Lave and her colleagues followed the shoppers—all selected because they were price conscious—around the store, observing them, taking copious notes,

---

[4] J. Lave, *Cognition in Practice: Mind, Mathematics, and Culture in Everyday Life*. Cambridge University Press, Cambridge, UK, 1988.

occasionally asking them to explain their reasoning out loud as they went about their shopping, and sometimes asking for explanations just after the transaction had been completed. Of course, this procedure is highly contrived. The very presence of an observer changes the experience of shopping. Thus, to some extent the study was not really one of people in their normal, everyday activities, but it was close enough for the purpose of the study, and moreover anthropologists have developed ways of going about such work so as to minimize any influences of their presence on their subjects' behavior.

Each of the researchers spent a total of about 40 hours with each of her or his subjects, including time spent interviewing them to determine their backgrounds (education, occupation, etc.). Though most of the shoppers were women, there were some men in the group. However, the researchers noticed no difference in the mathematical performances of the men and the women in the supermarket, so gender did not seem to be a significant factor.

Out of a total of approximately 800 individual purchases that the shoppers made in the course of the study, just over 200 involved some arithmetic—which the researchers defined to be "an occasion on which a shopper associated two or more numbers with one or more arithmetical operations: addition, subtraction, multiplication, or division." The shoppers varied enormously in the frequency with which they used mathematics. One shopper used none whatsoever, while three of the subjects performed calculations in making over half their purchases.

Among the arithmetical techniques that the researchers observed shoppers performing were estimation, rounding (e.g., to the nearest dollar or the nearest dollar and a half), and left-to-right calculation (as opposed to the right-to-left calculation taught in school). What seemed to be absent, however, were most of the techniques the shoppers had been taught in school. Lave and her colleagues set out to investigate where the school math had gone. In order to compare the shoppers' arithmetical performance in the supermarket with their ability to do "school math," the researchers designed a test to determine the latter. Again, the results were fascinating. Despite the significant efforts the researchers made to persuade the subjects that this was not like a school test, rather that its purpose was purely to ascertain what arithmetical ability they had retained since school and with nothing at stake, the shoppers treated it as if it were indeed a school test. They approached it in "math test mode," with all of the accompanying stresses and emotions.

Perhaps the shoppers' reaction was to be expected. After all, the "math test" did have all the elements of a typical school arithmetic test, including questions involving whole numbers (both positive and negative), fractions, decimals, addition, subtraction, multiplication, and division. On the other hand, the problems were designed to test the same arithmetical skills that the researchers had observed the shoppers using in the supermarket. For instance, having observed that shoppers frequently compared prices of competing products by comparing price-to-quantity ratios, the researchers included some problems to see how the subjects fared with abstract versions of such problems. For example, faced with an item costing $4 for a 3 oz. packet and a larger packet costing $7 for 6 oz., many shoppers would—in effect—compare the ratios 4/3 and 7/6 to see which was the larger. So the researchers would include on the test the question: "Circle the larger of 4/3 and 7/6." But the same shopper who did just fine in the supermarket would fail miserably on the school-type test. Overall, the shoppers' performance was rated at an average 98% correct in the supermarket compared to a mere 59% average on the test. Why?

One obvious difference was that the subjects approached the test questions as requiring a precise calculation, but they were much more likely to use estimation in tackling their real-life equivalents (although often with considerable accuracy). Another difference was that the shoppers in the supermarket were *not* using the arithmetical skills they learned in school. Rather, they were solving the problems another way. This conclusion is supported by the fact that performance on the test was higher the longer subjects had studied math at school and the more recently they had finished school, whereas neither length of schooling nor the time since schooling had any measurable effect on how well they did in the supermarket. Thus, if they teach anything, traditional school math classes seem to teach people how to perform on school math tests. They do not teach them how to solve real-life problems that involve math. A few schools have tried to address this issue by adopting a different approach that encourages students to think about problems and try to develop their own methods, but standard testing makes that a very difficult goal to achieve.

What seemed to make the biggest difference in test performance was the kind of test the shoppers were asked to take and the manner in which the questions were presented. This was shown by a further test the AMP researchers put the subjects through: a shopping simulation.

In their homes, the subjects were presented with simulated best-buy shopping problems, based on the very same best-buy problems the researchers had observed them resolve in the supermarket. In some of these simulations, the subjects were presented with actual cans, bottles, jars, and packets of various items taken from the supermarket and asked to decide which to buy among competing brands. In other simulations they were presented with the price and quantity information printed on cards. In this simulation, which was clearly a kind of "test" situation but with the questions of a shopping nature as opposed to school-like "math questions," the subjects scored an average of 93% correct.

The fact that the simulation was done in each subject's home, carried out by the researcher who had accompanied the subject on the shopping trip, seems to have been a significant factor. Not only did the subjects not view it as a "math test," they managed to approach most of the questions using the same mental resources they had used in the store. The researchers went to some effort to achieve this, such as giving the subjects the questions verbally in a conversational fashion and making frequent references to the actual shopping expedition the two had gone on together.

The importance of setting up the shopping simulation test this way becomes clear when you compare the AMP results with those from another shopping simulation test carried out by Deanna Kuhn.[5] Kuhn set up a table outside a Southern Californian supermarket, stopped customers about to enter to do their shopping, and asked them to calculate which of two bottles of garlic powder was the better buy, the 1.25 oz. bottle for 41 cents or the 2.37 oz. bottle for 77 cents, and similarly for two bottles of deodorant, one costing $1.36 for 8 oz., and the other $2.11 for 12 oz. The subjects were given a pencil and paper on which they could do their work.

The results were very different from those obtained in the AMP shopping simulation. Only 20% of the 50 shoppers who agreed to take the test were able to solve the garlic powder question, and not a great many more—just 32%—could solve the deodorant question. The enormous difference between the results observed in the AMP and Kuhn test procedures is almost certainly due to the way the subjects approached the two simulations. In the AMP simulation, the shoppers seemed to understand that they were to imagine they were actually shopping, whereas Kuhn's subjects seemed to view it as "taking a test." In fact, Kuhn's results were very similar to those obtained in the school-like tests administered to the AMP subjects. This confirms that you can carry out the test outside a supermarket and phrase the questions in terms of shopping, even to the point of presenting the subjects with actual items taken from the supermarket shelves, but if the subjects view it as a "math test," that is how they will approach it. As a result, they will struggle to use their long-forgotten—and possibly never fully understood—school math procedures. And more often than not, they will fail. These findings are of crucial importance in designing any video game intended to help students acquire mathematical thinking.

---

[5] Noel Capon and Deanna Kuhn, "Logical Reasoning in the Supermarket: Adult Females' Use of a Proportional Reasoning Strategy in an Everyday Context", *Developmental Psychology*, Volume 15, Issue 4, July 1979, pp. 450–452.

 # The Perfect Medium

There has to be a rational explanation for the huge disparity between the high performance level that ordinary people can achieve when they have to use "street mathematics" in everyday settings, and the terrible results many exhibit when faced with equivalent tasks in paper-and-pencil, symbolic form. Why do so many people claim "I just can't do math," when all the evidence indicates that they can do it just fine when they need to—provided it is not the school variety of math? Come to that, why *is* math so often presented using symbols, both in schools and elsewhere?

The Trouble with Symbols

If you ask someone to describe or to draw a picture of someone doing mathematics, almost certainly the image you will get is of someone writing symbols—on a piece of paper, a blackboard, or, if they are taking their cue from a number of portrayals of mathematicians in the movies or on television in recent years, on a window or a bathroom mirror. (Real mathematicians never do that, but it looks cool on the screen.) We identify doing math with writing symbols, often obscure symbols. Why? Because for the two and a half thousand years since the ancient Greeks initiated modern mathematics around 500 BCE, that was the only effective way to record mathematics and pass on mathematical knowledge to others.

Notice that I didn't say that was how people *do* math. If you think about it, *doing math* is something that takes place primarily in your head. What you write down when you are doing math are the facts you need to start with, perhaps the intermediate results along the way and, if you get far enough, the final answer at the end. But the *doing* math part is primarily a thinking process. And as the studies carried out in Recife and elsewhere demonstrate conclusively, even when people are asked to "show all their work," the symbolic stuff that they write down is not necessarily the same as what goes on in their heads when they do math correctly. In fact, people can become highly skilled at doing mental math and yet be hopeless at its symbolic representations.

So why do we still bother teaching symbolic math? There are some excellent reasons. One is that it is essential for modern life that there are enough people around who are really good at symbolic math. Science and engineering depend heavily on symbolic math. A second is that it is a key gateway to more advanced mathematics and a lot of other subjects as well. Still another is that the *process* of trying to master school algebra has proven scholastic benefits across the educational spectrum, regardless of how well the student does on the school algebra exam. But how often do ordinary people find themselves faced with having to solve a paper-and-pencil, symbolic math problem? Outside of school and college, and leaving aside those people who solve math problems for recreation—and there are a lot more of those people than is popularly assumed—modern life simply doesn't require such ability.

Except, of course, when it comes to passing tests. Though there is a lot wrong with the way the United States designs mathematics tests, society has to have measures of performance. At the moment, my focus is on designing video games to develop effective mathematical thinking, but video games can also play a role in improving test scores even in the United States, and I'll come to that in Chapter 12.

What modern life does require—unless you are content to live the life of a mollusk, not knowing or understanding what is going on around you or being vulnerable to being ripped off and misled at every opportunity—is a good working facility with numbers and quantity. If you think that a 10% price increase this week followed by a 10% price reduction next week will bring the price back to where it started, then sooner or later you are likely to be ripped off. (The final price will be 99% of the original.) If you think that a 50% chance of rain Saturday and a 50% chance of rain Sunday means a 100% chance of rain over the weekend, you are likely to be in for a surprise. (The correct answer is that the chance of rain over the weekend is 75%.) If you think that a newspaper headline saying that the rate of inflation is falling means things will become less expensive, you are in for a shock when prices continue to rise. (A decrease in the inflation *rate* means that prices are not going up as fast as before; but they can still be going up.) You can't read a newspaper or watch the TV news without being inundated with statistics, and that includes the sports reports.

But think about it for a minute. We're not talking here about doing math symbolically, school fashion. What modern life requires—and it requires it in spades—is the "in-the-head" stuff. The stuff that, as study after study has demonstrated, and which you may well have experienced yourself, almost anyone can pick up and get good at (or at least as good as they need to be).

So why do we try to teach people how to develop this useful, everyday ability by forcing them to learn and practice all those symbolic rules—a method that for most people demonstrably fails? Even when you come across someone who is able to ace all the symbolic math tests, and there seems to be at least one such person in every school math class, they are often unable to apply that ability in a real-world setting.

Teaching everyday math by way of symbolic mathematics denies mastery of the former—which the Recife study and others like it show is within everyone's grasp—to all those who fail to master the latter, which a great many do.

So why do we teach numerical mastery the way we do? Because for 2000 years, it has been the only way available at a system-wide level. If all teenagers could be put in a series of engaging, real-life situations that mattered to them, then like the Recife children, these students would eventually start to develop good, real-life number skills. But that approach simply is not practical.

Note that I am not saying we should abandon trying to teach algebra and focus entirely on mastering arithmetic in real contexts. We should teach both, for reasons I've mentioned. But acquisition of everyday number skills should not *depend* upon mastery of symbolic mathematics. On the contrary, mastery of number skills should precede the attempt to learn symbolic mathematics. To be sure, sufficient mastery of symbolic mathematics is a crucial part of achieving mastery of working with numbers. But the general ideas represented by the symbols need to be grounded in sufficient real-life, concrete experiences. Though all of this is well known, the current approach largely tends to approach these skills in the reverse order. How did we arrive at this situation?

Math for All

Until the beginning of the thirteenth century, mathematics was a specialist occupation. The European manufacturers and traders would employ people to do their mathematics for them. Those people would learn their skills by working alongside an expert as an apprentice, not unlike the young children in the marketplace in Recife. Much of what they did involved either finger arithmetic, a remarkable system that worked for numbers up to 10,000, or the use of mechanical aids such as counting boards—flat surfaces ruled with lines for place-value, on which the user moved "counters" or pebbles—and the various kinds of abacus. In 1202, a young Italian mathematician named Leonardo of Pisa (who a later historian would dub "Fibonacci") completed a book called *Liber abbaci* (spelled with two *b*'s, which makes the title translate as the "Book of Calculating") in which he described how to do calculations symbolically. It was one of the first teach-yourself arithmetic textbooks oriented to practical mathematics usage in the western world, and the one having by far the greatest impact on the spread of symbolic calculation in Europe. The methods Leonardo described had been developed in India several hundred years earlier and brought by Muslim traders to North Africa. Leonardo encountered them when, as a teenager, he joined his father who had been posted there as a Pisan customs official. The methods Leonardo recorded are essentially the same ones taught in school math classrooms ever since. You can see for yourself: an English language translation of Leonardo's book was finally published 800 years later in 2002.[1]

With the new approach to teaching math brought about by the completion of *Liber abbaci* (and literally hundreds of similar works that followed), it was no longer necessary to learn mathematics one-on-one as an apprentice. You could obtain a copy of the manuscript (and later a printed book) and teach yourself. Or a teacher could use the text to show a classroom full of students how to do it. What had until then been a specialist profession mastered by only a few suddenly became accessible to all.

But this democratization of mathematics came at a price. When most people solve a mathematical problem, they do so as a step in achieving something else— to manufacture something, to exchange currency, or to buy or sell a product. To use the methods described by Leonardo to solve a real-world problem, you first have to convert the problem into symbolic form, then apply the rules to solve the symbol problem, then translate the symbolic answer back into real-world terms. To help people master this part of the task, the infamous mathematical "word problems" (also known as "story problems") were developed. (*Liber abbaci* is full of them, and they can be found in much earlier mathematical works, such as ancient Babylonian tablets and Egyptian papyri!) But as generations of students have discovered, it takes a huge amount of effort to master word problems, and many never get the hang of them. The human brain simply is not well suited to do that kind of thing. (I explain why in my book *The Math Gene*.)

Still, the approach Leonardo described was the only game in town. And so, for 800 years since the appearance of *Liber abbaci*, that is how arithmetic has been taught. Geometry and algebra have been taught using textbooks for even longer—Euclid's famous multi-volume geometry and algebra textbook *Elements*, written around 350 BCE, looks much the same as any present day geometry or algebra textbook. When you have only one method to teach an important and useful skill—in the case of interest to us, real-world arithmetic—to a large number of people, then that method is what you use, and everyone has to struggle through the best they can.

---

[1] Laurence Sigler, *Fibonacci's Liber Abaci*, Springer-Verlag, New York, 2002.

That makes it doable for those with sufficient time and motivation, but it does not make it easy. By comparison, suppose the only way to learn to play or appreciate music was by learning to read musical notation. Few of us would become musicians and many people would never learn to appreciate it. But that is not what happens. You can play, appreciate, and even compose music without knowing how to read the notation. Before the days of musical recording technologies, musical notation was the only way to store music and to distribute it widely, apart from passing it from person to person. People who wanted to play some new music they had not heard before, and who did not have someone on hand who knew it and could sit down alongside them and teach them, had no alternative but to learn to read the notation. The point is simply that the notation is not the music; it's just a way to represent it for storage and distribution—and it's a very unnatural (though effective) way of doing so.

It's the same with mathematics, at least the kind of math that comes up all the time in the everyday world—the math that everyone needs, or that at least enables them to do much better in life if they can do it. I will refer to this collection of abilities as *everyday mathematics*. (*Basic math* is another commonly used term for the same thing.) Generally speaking, everyday mathematics includes the following topics: basic number concepts, elementary arithmetic, arithmetical relationships, multiplicative and proportional reasoning, numerical estimation, elementary plane geometry, basic coordinate geometry, elementary school algebra, quantitative reasoning, basic probability and statistical thinking, logical thinking, algorithm use, problem formation (modeling), problem solving, and sound calculator use.

Today's world requires a degree of mastery of each of these mathematical topics. "Mastery" in this context means not just being able to perform calculations with fluency. It is also important to have a good conceptual understanding of numbers, arithmetic, and reasoning, particularly in the context of real-world applications.

Let me stress that references to "basic skills" (or just "skills"), "conceptual understanding," and "mathematical thinking" are not meant to imply that these are distinct categories of mental activities. These are suggestive and helpful terms, but they defy precise definition and there is a lot of overlap among them. They are also not entirely independent of one another. In particular, a person can have computational skills without much conceptual understanding, though that invariably leads to problems if any further progress in mathematics is required—and many people find that it is.

For example, a child can learn the multiplication tables by rote to the point of rapid, automatic recall, without having any understanding of what multiplication is or how it relates to things in the world. Indeed, many successful adults never fully understand multiplication, though they can use it correctly in certain circumstances. This is demonstrated dramatically by the widespread belief that multiplication of positive whole numbers is repeated addition. That this is not the case is easy to explain, though it is evidently not so easily understood.

Addition applies to objects of the *same type*, measured using the same units. Mathematically, you cannot add apples to oranges. Five apples plus seven oranges is just that. Addition requires you to first reclassify the items so they are of the same type—in this case, fruits—so you can say "Five fruits plus seven fruits gives twelve fruits."

In multiplication, however, the two things being multiplied must necessarily be of *different types*. You can say, "If I have five bags each containing seven apples, then I have 35 apples." But note the units: five is counting the number of bags, seven is measuring apples per bag (i.e., it's a ratio) and 35 is counting apples. In terms of the units, BAGS × APPLES PER BAG = APPLES. It is true that if you empty out all the bags, then you can do a repeated addition and you will get the same number, 35. So the numerical answer is the same. But that does not make the operations the same.

In fact, there is an even more fundamental difference between addition and multiplication when using them in the world. For addition, the order in which the quantities appear does not matter; five apples plus seven apples is (conceptually and numerically) the same as seven apples plus five apples. But for multiplication the order in which the quantities appear is significant; five bags of seven apples is not the same as seven bags of five apples, either as a scene to look at, or conceptually, or in terms of the units associated with each number. On the other hand, when you move to a totally abstract representation, free of units, it is indeed the case that $5 \times 7 = 7 \times 5$. (This is an instance of the commutative law.)[2]

In fact, a better way to approach multiplication is to think of it as *scaling* one quantity by another. This has the clear advantage of working for fractions and real numbers, where the flawed "repeated addition" notion fails. But even that does not tell the complete multiplication story. For a really good treatment of the basic arithmetic operations and the difficulties children have fully grasping them on a conceptual level, see the book *Children Doing Mathematics (Understanding Children's Worlds)* by Terezinha Nunes and Peter Bryant.[3]

When something as basic as multiplication of positive whole numbers turns out to be conceptually fairly complex, you know that everyday math involves a whole lot more than rote learning of a few facts. You can learn to calculate with numbers without any real understanding of the underlying concepts. But applying arithmetic to things in the world, to quantities, and understanding the relationships between those quantities, requires considerable understanding of

---

[2] At the abstract level, "repeated addition" is not even well defined, let alone equal to multiplication or anything else, but this is a subtle point I won't pursue here.
[3] Wiley, New York, 1996.

those underlying concepts. Once that understanding has been achieved, however, the calculation skill really does, as I claimed earlier, "come for free."

Eleven Principles for an Ideal Learning Environment

For two-and-a-half thousand years, we have used symbolic mathematics notation because it was the only way we had to store and widely distribute mathematics—even basic, everyday mathematics. But suppose you had complete freedom to design an alternative method of storage and distribution for everyday mathematics, from which people could learn how to do it. One that you hope would produce that near perfect 98% accuracy performance exhibited by the young market traders in Recife. Such a storage/distribution device would be a mathematical equivalent to the iPod for storing and distributing music so anyone could listen to it and appreciate it. What properties would your storage/distribution device need to have? (Yes, you know where I am going with this; but it's a useful exercise to put those expectations to one side and focus entirely on the pedagogic requirements for good learning of everyday mathematics.)

Principle 1
It would have to be a "real-world environment."

This is the main lesson from the studies carried out in Recife and elsewhere. To hit that 98% accuracy rate, the learning environment needs to be sufficiently like the real world situations where ordinary people use mathematics—the market, shops, offices, factories, laboratories (this one may be a bit specialized), sports facilities, recreational areas (such as mountains or national parks where people use navigational skills that depend on trigonometry and arithmetic), etc. The mathematics to be learned has to arise naturally in that environment, and have meaning in it, and the learner in that environment has to be motivated to carry out the tasks that involve that mathematics.

The key condition in that last paragraph is that the learning environment should be sufficiently like the applications in which the mathematics will be used. What the Recife study and others like it show is that, in real world settings where people deal with math, it's not abstract, symbolic math that they use; rather, what they do is reason about the real world. The young stallholders in the Recife street market were not reasoning about numbers or "doing arithmetic" in the school sense; they were selling things, such as coconuts, fruit, vegetables, straw hats, and cheap plastic sandals, and they were handling money. When people are in a real-world setting, faced with reasoning about (real) things in that setting, they do just fine, even if the reasoning is about numbers of things, measurements of things, etc. It's decontextualization and abstraction that cause many people problems, and that's why they have difficulties with symbolic math. For effective math learning, people need to be in either a real-world environment or one sufficiently like the real world so that it feels like being in the real world. For example, the environment could be a fake marketplace set up in the school playground. (Remember the subjects in Jean Lave's Adult Math Project, who scored a respectable 93% when asked questions posed and presented in a fashion that evoked the reasoning skills they used when doing real shopping.) Or it could be an environment having very little in common with the material world, provided the learner could be cognitively immersed in it.

One other feature that was clearly present in the Recife case is that for the child market stall holders, what they were doing really mattered to them; they had a vested interest in the outcome. This was the case for the subjects in the other studies I've cited. Motivation is a huge factor in mathematical performance. If motivation alone were sufficient, however, far more high-school children would do better on the SAT mathematics exam, because the students' future education and subsequent career can ride upon the result. The difference is that, for most children taking the SAT, it is merely the result of the test that matters, not the activities on the test, which are often unrelated to the child's scholastic ambitions (or at least the child perceives that to be the case). In other words, the SAT is perceived as merely a hurdle to be gotten over as well as possible. For the children in the Recife market, in contrast, the mental calculations they performed were an integral part of an activity that both had meaning for them and mattered to them.

Principle 2

The learning environment should be as similar as possible to the environment in which people will use what they learn.

A primary aim of basic mathematics education, the one that is relevant to most people, and the one I am currently focusing on, is to use it in the real world. (Another aim is to develop the reasoning ability that is required for the mastery of further mathematics, some of which can also be used directly in real world situations.)The best way to learn how to play tennis is on a tennis court—or, these days (and this is getting at the heart of this book), using a tennis simulator such as a Wii—not by reading about tennis and answering questions about how to play. The best way to learn to drive a car is to get behind the wheel and drive. When pilots learn to fly planes, they start out either in a real plane with an instructor alongside, or else in a simulator that is designed to provide them with a realistic experience of flying a plane. Why treat everyday math any different? (I should stress again that my focus here is everyday math skills. Symbolic, abstract math is something entirely different.) As the Adult Math Project showed, there is a performance cost associated with disparity between the learning environment and any application environment, but if those two environments are sufficiently similar, the loss in performance can be fairly minimal. Exactly in what way they need to be similar is not always clear.

It important to keep in mind though, that for mathematics the goal is for the learner to be able to use it in many different situations, so the learning environment should provide sufficient variation in circumstances of use so that the student can apply it in novel situations, not just those identical to the ones encountered during the learning process.

Principle 3

It is important to provide unlimited numbers of learners with exactly the same learning environment.

We are not seeking to provide each person with one-on-one mathematics education; rather, the aim is to provide mathematics education for everyone in the community. Social justice requires no less. One-on-one tutoring in real world environments (apprentice learning) is highly effective, but hugely expensive in terms of people's time. Moreover, since not all teachers have the same knowledge, ability, and teaching skills, personal tutoring inevitably entails a huge degree of inequality, which our society will not knowingly tolerate within the public education arena. For all their weaknesses, textbooks do in principle provide all students with exactly the same instruction. (Of course, when a textbook is used by a teacher, inequalities inevitably creep in, even with a good teacher, but the textbook does provide a leveling force, and many a determined student has overcome the handicap of a poor teacher by teaching him or herself directly from the book.)

Principle 4

The learning environment should be designed to allow a learner to repeat an experience facilitating a post mortem analysis of what took place.

One disadvantage of teaching someone mathematics in a real-world environment such as a market stall is that if the student gets something wrong, it's generally not possible to repeat the same problem and analyze the student's work as you go through the same steps a second time. That is a loss. We learn a lot from our mistakes, and educationally it is extremely effective to be able to repeat an experience where things went wrong and try to get it right second time.

Principle 5

The environment should be designed to facilitate variation along a single dimension or a given selection of dimensions.

Unless you are a mathematics educator, you might not think of this requirement. One of the most effective means for someone to learn a new mathematical skill is to see what happens when only one feature (perhaps one numerical variable) changes or a particular set of features changes, while everything else about the problem remains constant. In fact, some mathematics educators suggest that being able to compare scenarios that differ in just one particular or a selected group of particulars is the single most effective way to learn. This can be hard to arrange in a real-world environment, of course, where it may not be possible to change just one parameter without changing other features at the same time. But controlled variation in scenarios is effective, so let's include it in our wish list.

Principle 6

There should be some uniform means of assessing students' performance.

An obvious challenge with any kind of one-on-one instruction, particularly if the instruction is in environments that differ from one student to the next (which is the case for any instruction in real-life environments), is the difficulty of providing an assessment tool that is fair to all.

Some educators, dismayed by the difficulties and inequalities of testing, have even gone as far as to advocate the abolition of standardized tests altogether. But then the only way to provide reliable and effective education is to have one-on-one or small-group learning, where the instructor can measure the performance of each student individually. And even that has an unavoidable subjective element. For public education, regular, standardized assessment is essential. But does it have to have all the negative connotations and consequences we have come to associate with standardized tests? Perhaps we could find or create a learning environment in which standardized, regular testing seemed perfectly normal, or even enjoyable. You don't think that's likely? Fair enough, but I want to put it in anyway, just in case.

Principle 7

The environment should store and present the students with pre-planned learning experiences, some of them in a particular order.

One thing that good textbooks do well is to provide students with well planned learning experiences. Moreover, they present all students with the same experiences. An alternative learning medium that aims to do better than textbooks should do the same.

This should not be taken to mean recipe-book type instruction, where the student simply has to learn to follow the rules in a pre-set, unreflective, mechanical fashion. The student should still have freedom to navigate through the planned experience. By analogy, when you drive your car, you are constrained to follow the roads the highway planners have provided for you, but you have a lot of choice as to where to go—hence the phrase "the freedom of the open road."

Principle 8

It should be possible for the student to explore new concepts and practice new techniques at his or her own pace.

It takes most people time to assimilate new concepts or master new techniques, and the amount of time can vary enormously from one individual to the next. All of the great mathematicians throughout history spent many hours exploring concepts and ideas, "playing with ideas," trying one thing then another, looking for ways to connect the new and unfamiliar with the known and familiar. You wouldn't know this from reading their published work, which generally conveys the impression that it is the product of some superhuman intelligence that magically knows exactly which step to perform next. In mathematics, as in sausage making, the final result is made public but the process that leads to it is kept hidden. In the case of sausage manufacture, that's probably a good thing, but for mathematics it gives students the impression that, if they can't "see the light" right away, they simply don't have what it takes to do math.

Many teachers have suggested that time spent in individual exploration is the single biggest factor that differentiates students who are successful in mathematics from those who forever struggle. In particular, exploration time can outweigh to a considerable extent any differences in innate mental power.

Principle 9

The student should be given immediate positive (and ideally public) feedback for any success that is commensurate with that student's current level of attainment.

Though I have expressed this requirement in terms of the learning process, the environment should allow this to happen. We all like to be acknowledged when we have done something well. The acknowledgement should never be too small that we feel cheated, but equally it should not be so great that we feel we do not deserve it—after all, success in meeting the challenge is itself a reward. And while there is no doubt that positive feedback is valuable for the learner if received in private, it is a much more effective form of encouragement if done in public. Good teachers know this and use it often. It is one of the factors that can make multiplayer games and social network games better learning environments than single-user video games.

Principle 10

There should be sufficient "cost" to getting something wrong to motivate correction, but not so great that it leads to the student losing heart and giving up.

This too is a guiding principle for the learning process, but again the environment needs to allow it to happen. I touched on this factor in my final comments about principle 1; for the children staffing the market stalls in Recife, getting the math right really mattered to them—this was their families' livelihood. Unless it matters to a student whether he or she knows a certain fact or can carry out a certain procedure, it is unlikely that he or she will invest sufficient effort into achieving such knowledge or ability. At any one time, there are often several things each one of us would prefer to be doing than the task before us. As a result, mattering matters.

It should be evident that mastery of a particular mathematical technique can carry a "sufficient cost" to a student only if the student is already motivated to succeed in the overall learning enterprise that is taking place in that particular learning environment. Indeed, the greater the motivation to succeed in the overall endeavor, the more the student is likely to respond positively to greater losses resulting from failure—at least up to a certain threshold which likely varies from one individual to the next.

Principle 11
The learning environment should, if possible, provide an enjoyable and stimulating experience.

Learning mathematics is not easy—though traditional teaching methods have made it much harder than it needs to be. So while it does not have to be fun, making it as enjoyable as possible can enhance the experience. It should also be stimulating. As it happens, the learning medium that this book focuses on is both enjoyable and stimulating, characteristics that most mathematics textbooks do not provide.

The Perfect Medium

Today, we have an ideal medium for teaching what I am calling everyday math. We can create a learning environment that has all of the 11 desirable principles listed above. We have been building such environments for about 30 years now, but until recently the development costs were prohibitive. What has driven down the costs to the point where it is now feasible to make use of these environments for educational purposes is that they have been adopted and developed by a highly lucrative business. I am referring, of course, to the virtual worlds utilized by the video game industry. Those virtual worlds can be represented on a computer screen in either an explicitly two-dimensional (2D) or simulated three-dimensional (3D) fashion, or in an intermediate form known as 2.5D. The important feature is that they are immersive; the user has a strong sense of *being in a world* and *acting in it*.

I am not referring here to video games as such, but rather the virtual worlds they simulate and present on the screen. It will be important to bear this distinction in mind, because much of my focus will be on the value of video games for basic math education. Some of the 11 principles listed above pertain purely to the environment, while others seem to argue for, in addition, a particular activity structure within that environment, for which a game might be the most natural vehicle.

In fact, this book actually argues for *two* innovations in basic math education. The first is to take advantage of the power and unique suitability of virtual worlds to facilitate situated learning in simulations of the real world (or imaginary variants of the real world). The second is game mechanics. It should not come as a great surprise that game mechanics can play a role in learning, since they fit naturally into those environments. A game can provide a structure for the learning that takes place in the environment—the curriculum, if you will. Game mechanics also provide significant motivation to persevere.

An analogy to books may help. Books are used for several different purposes. One use is for learning mathematics, in which case we call the book a textbook. A book can also be a vehicle for telling a story, and then we call the book a novel. Same medium, different purposes. Likewise, a virtual world can be used to teach mathematics using a traditional pedagogy or as the setting for a game that facilitates learning.

To pursue this analogy a bit further, writers of successful textbooks know that a powerful learning effect can be achieved by borrowing as much as possible from novels—by telling a story. There are significant limitations to how much this can be done, but it is possible and I have done it myself in some of my textbooks. In the case of virtual worlds, however, there are far greater opportunities to use elements of game mechanics in the teaching process. The reason is that the learning that takes place in such an environment is experiential—the student learns by doing. This, in a nutshell, is why this book is primarily about video games played in virtual worlds, and not just the use of those worlds as an environment in which to teach math in a more traditional fashion.

Having laid my cards on the table now, it's time to take a look at the educational medium I am advocating. Since I anticipate that many of my readers are not familiar with video games, I will start at the beginning, and experienced gamers can skip briefly though the remainder of this chapter.

Game Genres

Twenty years from now, a section such as this will not be necessary. By then, most parents, teachers, and education leaders will themselves have grown up playing video games. They will recognize and be familiar with the features of good video games that make the medium ideal for developing mathematical skills and understanding. But right now we are in the transition period where many parents, teachers, and education leaders are not at all familiar with video games, while almost all of the students are—97% according to a December 2008 study by the Pew Research Center. This is not to say that gamers are predominantly school-aged. They are not, and I will look at the current demographics of the gaming community presently.

Thus, one audience I am trying to reach with this book comprises all those non-gamers with an interest in mathematics education. That would include all non-gaming parents, teachers, professors of mathematical education, and education leaders. To those readers, let me say that, *before you try to read further in this book, you should go out and get a good video game—and play it!* If you don't do that, you may have a difficult time following what I say. In either event, let me sharpen the focus on the kind of video game I am talking about.

To the casual observer, all video games look much the same, but there are in fact several different genres of video games. Forget the popular misconception that they are all about fast-reaction thumb action. Although some are, even most so-called "fast-action games" are rarely exclusively so; rather, they generally involve considerable strategy formation and problem solving prior to engaging in a fast-and-furious, real-time battle. Many "conflict" games don't require fast action by the player at all, and are almost entirely about problem solving and strategy development, with the on-screen conflicts essentially resolved within the computer before the battle commences.

At present, hardware differences distinguish two different kinds of games. Some games are played on special consoles such as the Microsoft Xbox or the Sony PlayStation 2, and others run on ordinary personal computers (PC). To over-generalize somewhat, the former tend to be action-based and require fast responses with a joystick and various button controls, while the PC games are primarily strategy-based fantasy games in which the player has lots of time to plan out the next action. For the most part, the action-based console games derive from video arcade games and the PC fantasy games trace their lineage to the non-computer role-playing game *Dungeons & Dragons*. But these days, console games increasingly present a significant intellectual challenge and many fantasy games include some real-time keyboard action where speedy reaction is important. As the game industry continues to develop, the current platform distinction is likely to disappear, with the emergence of a single form of universal game playing device with various specialized controls for specific kinds of games, such as the Nintendo *Wii* or Microsoft's *Kinect*, which require whole-body motion by the player.

In the early days of computer games, the screen representations were mostly two-dimensional. In many games today, the player enters (through his or her in-game character) a fully immersive three-dimensional environment. This provides the player with a degree of "presence" that can only be appreciated by playing such a game—by literally *entering* the simulated world offered by the game.

Some video games are played alone, some with or against other players in the same room, and in some games many players (perhaps thousands) come together in common game spaces over the Internet. Some video games are overtly competitive, either a player against a digital opponent that is part of the game software, or player against player; others focus on social or civic life and community building. In some games the player sees the simulated computer world as if through his or her own eyes, such as the so-called "first-person shooter" games; in others the player controls an in-game character that is visible on the screen at all times. Many games involve combinations of two or more of these elements. As I've indicated, my primary focus, at least initially, is on games in 2D or 3D virtual worlds played on personal computers using the mouse and keyboard to direct the action.

In many such games, the player creates a character (an "avatar") to represent him- or herself, and controls the actions of that character in a simulated fantasy world. In the game world, the player's avatar interacts with digital characters controlled by the game program (called "non-player characters" or NPCs) and also with the characters of other human players. As of this writing, one of the most successful games of this kind is the MMO *World of Warcraft*, with 15 million subscribers worldwide and rising. The name may be misleading; think J. R. R. Tolkien's fantasy novel *Hobbit*. If you are unfamiliar with MMOs (or video games in general), that game would be my number one recommendation for you to try right now. Prior to the appearance of *World of Warcraft* in November 2004, the most popular game of this genre in the United States was *EverQuest*, with around 350,000 players. The advantage of choosing an MMO as your example is that they have all the features that educators can take advantage of. Because MMOs are time consuming and expensive to produce, the majority of educational video games we are likely to see in the near future will have only some of the features of an MMO.

Let me repeat my previous advice with a more specific recommendation. Before you read further in this book, you should get hold of *World of Warcraft* and experience game play to a level of reasonable proficiency, say, Level 10. (The game has 60 levels—80 if you buy the two expansion packs that came out in early 2007 and late 2008—and gets progressively more difficult the higher you go.) If you have never played a video game before, you will be surprised how cognitively challenging it is, even at the lowest levels. Getting some initial help from a gamer— friend, child, grandchild, whoever—can be a real help until you get the hang of the basic mechanics of such games. Without some first-hand experience playing such a game, there is really little chance you will fully understand or appreciate much of what I say in this book.

The Myths and Realities of Video Games

Despite my urging that you put down this book and start playing a good video game, I suspect that many of you ignored that advice—at least for now—and kept reading anyway. Reading is, after all, what we non-gamers (and I was one until relatively recently) know from experience to provide a highly efficient way of acquiring new knowledge. Why waste time playing games, especially if it requires some time and effort to set up in the first place? That question is, of course, rhetorical. But let me ask one that is not. Is playing video games a waste of time? Certainly some parents and teachers think so. But if they have never themselves played video games, how do they know they are a waste of time? Well, they will say (and I know this because I've asked them), "It's *obvious* they are a waste of time," or "*Everyone knows* they are a waste of time." Those responses hardly count as scientific. That such a conclusion can be "obvious" is highly dubious when expressed by someone who has never played a video game, and "everyone" presumably refers to everyone who has not played video games. Where is the evidence? Where are the scientific studies of the dangers, if any, of playing video games?

In fact, there have been studies of video game players, but what they have established is that video games, far from being a negative force in modern society, offer various benefits. To be sure, the news is not all reassuring. There is some evidence that excessive playing of violent video games correlates with the tolerance of real-life violence and the likelihood of acting violently, and it is possible, though yet to be established definitively, that the former causes the latter. There have also been cases of video game addiction, though not nearly to the degree that the popular media would have you believe. But the focus in this book is not on violent games, nor does learning math through a video game require excessive playing time. Would there be a public outcry, I wonder, if use of a mathematics educational video game led some players to become addicted to doing math?

One obvious benefit of video games—and this one surely *is* obvious—is that they provide entertainment. In that regard, they are like novels, plays, movies, television, concerts, sports, and other recreational activities. There is nothing wrong with that. "Ah," you may say, "but those other forms of entertainment are much better for people. For instance, reading novels also exercises the imagination and has educational benefits, and engaging in sports is good for health and general fitness."

Well, playing video games also exercises the imagination and has educational benefits that I will show. True, playing most video games does not provide any physical exercise, but then nor do reading or going to a play or a concert or any of many other things people typically regard as valuable leisure time activities. "Yes," you may say, "but video games take up so much time, they prevent people doing all those other valuable things." This comment ignores the fact that the same could be said of reading or listening to music. But in any case, the data—and there is a lot of it—indicates exactly the opposite. Video game players, as a group, actually spend more of their time doing other things than non-players!

That popular media image of the video game player as an antisocial male teenager is about as far from reality as it could be. True, there have been and continue to be cases of compulsive video game addiction. But then, a small proportion of the population always seem to be susceptible to addiction to one thing or other—food, alcohol, and nicotine being three of the more prevalent. The issue is the psychological tendency toward addiction, not the object of that addiction. In any case, it's time to hit the myth with some hard data. First, notice that in general it's the most exceptional cases that make the news. Video games are no different. But in evaluating worth, it is important not to confuse the exceptional, excessive behavior with the norm. With any human leisure activity, whether playing golf, enjoying alcohol, eating, or watching television, there will always be a small minority of individuals who carry it to excess and become addicted. The psychological reasons are complex and varied, but often these are people who may be seeking some form of escape from certain things in their everyday lives. While it's also true that some video games are extremely unpleasant, the same is true of some novels and movies. In any case, we don't need to base our opinions on sensational news stories.

Because video games are fairly new, make use of the latest technology, and involve large amounts of money, there have been several scientific studies of gamers and of game playing. For instance, the highly regarded, independent Pew Research Center tracks video game play fairly closely and produces regular reports. Their December 2008 report provides the following data:[4]

- 53% of American adults age 18 and older play video games.
- Roughly one in five adults (21%) play everyday or almost everyday.
- 97% of teens play video games.
- 81% of 18-29 year olds play games, while only 23% of people over 65 do.
- Men (55%) are slightly more likely than women (50%) to play video games.
- Urbanites (56%) are a bit more likely than rural dwellers (47%) to play.

Video game playing also correlates positively with educational level and activity:

- 57% of people with some college education play video games, significantly more than high-school graduates (51%) and those who have less than a high-school education (40%).
- 76% of current students (82% of full-time and 69% of part-time students) play video games, compared with 49% of non-students.

So much for the negative popular image!

As for the myth that playing video games has little or no educational value, I cannot really do any better than refer you to two excellent books on the topic. My first recommendation, John Beck and Mitchell Wade's book *Got Game: How the*

---

[4] The Pew report is based on a survey of 1,102 teens (1,064 teen gamers) conducted November 2007-February 2008, with a margin of error of ±3%. For more information about teens and gaming, see "Teens, Video Games and Civics" (Pew Internet & American Life Project, September 2008) Available at: http://pewinternet.org/PPF/r/263/report_display.asp. The Pew researchers note that only 21% of gamer teens report playing MMOs, which, as a "Johnny-come-lately" among game genres, are still relatively early in their growth curve.

Gamer Generation Is Reshaping Business Forever,[5] is aimed at the business world. It focuses on the social, collaborative, and managerial skills many games develop. Those skills include problem solving, resource management, decision making under uncertainty, decision making under pressure, multitasking, and (particularly the MMOs) interpersonal skills and organizational ability.

The United States Army was one of the first large organizations to spot these positive learning outcomes from video games, and has invested many millions of dollars in their development and use. Their game *America's Army*, a first-person shooter (the name reflects the player's viewpoint and what the game involves) that runs on the Microsoft Xbox, was developed to boost recruiting and has over 8 million registered users. Many large corporations also use video games to train their managers.

Beck and Wade list a number of characteristics of individuals who have grown up playing video games. This group comprises the majority of current school- and university-aged children and young adults. Beck and Wade's list includes:

1. failure doesn't hurt,
2. risk is part of the game,
3. feedback needs to be immediate,
4. used to being the "star,"
5. trial and error is almost always the best plan,
6. there's always an answer,
7. I can figure it out,
8. competition is fun and familiar,
9. bosses and rules are less important, and
10. used to group action and conflict.

Most leading educational thinkers would argue that all but one of these are precisely the characteristics of a good learner and a desirable learning environment. The one exception would surely be number 5, but I think that is more a feature of the simplistic way the authors state the principle. Successful game play almost always requires *informed* trial and error, where the player learns from previous experience in the game what actions to avoid and what is best to try first. It is possible to argue that number 6 is also not in tune with the real world, including real-world applications of mathematics. But it is a common feature of our K-12 educational system, particularly mathematics education, to confine teaching tools to problems that have answers.

Moving on to my second book recommendation, James Paul Gee, a professor of education at Arizona State University (he was at the University of Wisconsin

---

[5] Harvard Business Press, 2004.

when he wrote the book), uses the title of his book to make clear his main focus: *What Video Games Have to Teach Us About Learning and Literacy.*[6] Gee points out that the features required for a video game to succeed in attracting and keeping users are essentially key features of education, and therefore that video games are in a sense all about learning. Gee acknowledges that to people who typically think of learning as something that takes place in classrooms or through reading at a desk, playing video games does not *look* much like learning, and much of what is learned is not what constitutes the main focus of a typical school curriculum. But Gee asserts that such an appearance does not mean that game players are not learning. As his book title suggests, Gee's main focus is literacy, but he means that in a very general sense of not simply being able to read, but the ability to become familiar with the entire culture in which a particular text is embedded.

Gee lists 36 key principles of education that go into the design of a successful video game, and that helps open the door to the design of future video games that are targeted at specific content learning. I'll look at all 36 of Gee's principles in detail in Chapter 8, but by way of illustration, here are two:

Achievement Principle
(Gee's Principle 11)
For learners of all levels of skill there are intrinsic rewards from the beginning, customized to each learner's level, effort, and growing mastery and signaling the learner's ongoing achievements.

Practice Principle
(Gee's Principle 12)
Learners get lots and lots of practice in a context where the practice is not boring (i.e., in a virtual world that is compelling to learners on their own terms and where the learners experience ongoing success). They spend lots of time on task.

Teachers will recognize these at once as among the basic principles of good instructional practice.

As Gee points out, game developers did not set out to incorporate his 36 educational principles into their games. Indeed, many—though not all—game de-

---

[6] Palgrave Macmillan, 2003.

velopers to date have most likely been completely unaware of educational theory and practice. In fact, many successful game developers are college dropouts who left formal education to develop games. But in order to be successful, their games had to exhibit the educational features Gee identifies. Many of those principles apply to games designed to teach mathematics.

Having at least laid the groundwork for making the case that video game technology is extremely well suited to teaching basic mathematics, it's time to take a closer look at the kind of games that I have in mind, and the immersive environments in which they are situated. I'll base my discussion on MMOs, taking *World of Warcraft* as my main example, but practically everything I will say applies to any multiplayer game played in a virtual world. In fact many of the educational benefits of video games can be achieved with single-player games and with games having fairly simple 2D representations. As I noted before, MMOs simply offer all the features in one package.

One reason for using *WoW* for my illustrations is personal: it is my favorite game, and the one I understand the best. More significantly, however, there have been many studies of educational uses of MMOs, and almost all use *WoW* as their main example, so doing so here facilitates comparisons. One such study, which I drew upon in writing this book, and which I recommend highly, is Michele D. Dickey's scholarly article, "Game Design and Learning: A Conjectural Analysis of How Massively Multiplayer Online Role-Playing Games (MMORPGs) Foster Intrinsic Motivation."[7]

Having decided to use the all-encompassing MMO as my model, however, I should point out that such games are highly complex—and hence difficult to design and build—and thus the MMO genre is unlikely to be the one chosen by most developers of math ed video games.

Games without End

MMOs such as *World of Warcraft*, while clearly entertainment media, are not competitive games, in the sense of having an ending with winners and losers. Rather, they are fantasy worlds that players can explore and experience indefinitely.[8] Some players choose characters whose primary function is combat, such as my human warrior character in *World of Warcraft*, and for them a lot of the time spent in the game environment does involve competition, either against game characters (the NPCs) or other human players. For such players, the experience has many

---

[7] *Education Technology Research & Development*, 55 (3), 2007, pp. 253–273.
[8] They are "infinite games" in the terminology of James P Carse's excellent little book *Finite and Infinite Games: A Vision of Life as Play and Possibility* (Ballantine Books, 1987), a philosophical book with religious connotations that was not written with video games in mind, but that everyone interested in video games should read.

Figure 3. My two characters (avatars) in *World of Warcraft*.

elements of competitive games—a more accurate description might be that the game is an open-ended tournament of competitive encounters. But other players choose very different characters—priests, magicians, healers, tradespeople, etc.—and for them the experience is different.

One feature of MMOs that distinguishes them from books, plays, movies, or sporting games is that they are open-ended. The presence of thousands of other players to interact with means that there are always new experiences to enjoy. In addition, the game environment itself changes as the game developers continually add new regions and introduce new challenges. Edward Castronova, one of the few but growing number of academics who have made an in-depth study of such systems, refers to them not as games but *synthetic worlds*, a term I find particularly appropriate.[9]

For the benefit of readers who are not familiar with these games, let me describe briefly what playing such a game involves. The first thing a beginning player encounters (after digitally signing a long user agreement,) is to create their character, the *avatar* that will represent them in the game world. The game designer will provide many options to chose from: gender, class of character (games typically involve two or more "races", some on friendly terms, others sworn enemies), type of character with corresponding capabilities (e.g., warrior, priest, hunter, rogue, warlock), skin type and coloration, dress, and equipment. This allows players to build characters that suit their personalities and which they feel comfortable with as projections of themselves. Experienced players often have several characters of different kinds, providing different game experiences within

---

[9] See Edward Castronova, *Synthetic Worlds: The Business and Culture of Online Games*, University of Chicago Press, Chicago, 2005.

the same game environment. According to research by Dr. Nick Yee,[10] roughly one third of players online at any time are playing a character of the opposite gender to their own. Players' characters encounter and may interact with both other players' characters and with the NPCs, the avatars that are controlled by the artificial intelligence built into the game. Figure 3 shows my two avatars in *World of Warcraft*.

The underlying game dynamics include exploration, meeting challenges, solving problems, advancement (these games have achievement "levels," each of which confers benefits on the player's avatar), acquiring items and making things with them, earning money, and buying and selling items—all very much like real life, in fact. "Leveling," the process of increasing the "seniority" of a character in the game, turns out to be an extremely powerful motivator in these games. It is rare for a player who, finding him or herself within a few challenges of "leveling up"—i.e., moving to the next level—as midnight approaches, will stop playing and go to bed. That goes for 50-plus-year-old university professors with a 9:00 AM class the next morning as much as for their students! Leveling up provides a powerful incentive for mastering skills, which for players of a math ed video game will include mathematical thinking.

Once a player has chosen his or her character, the basic characteristics of that character remain fixed, but one of the purposes of game play is for a player to develop that character, performing actions that enable it to acquire new skills, different clothing, various kinds of equipment, wealth, perhaps property or means of transportation, etc. If that sounds a lot like real life, then that indeed seems to be much of the appeal of multiplayer games in virtual worlds, which offer many of the challenges, motivations, experiences, and rewards of real life, but in an often fantastical fantasy world, and free of the sometimes painful and long lasting consequences of making mistakes that the real world entails.

Once the beginning player has selected his or her character, a short movie (known as a "cut sequence") plays that describes the "historical" background for the activity that will follow, and explains in particular the history and culture of the chosen character. This is not unlike the initial experience of a student entering a university, who will attend an orientation session describing the history and culture of the institution and setting out what will be expected of him or her.

Then the action starts. The player controls his or her character using simple mouse and keyboard actions, directing the character to move in different ways and perform various actions. The player can generally choose any view of the world he or she wants: from behind the character, face on to the character, from directly above or directly below, and can change the view at any time by using the mouse to move the "camera" continuously through a full spherical geometry. The player can also choose to view the world directly through the eyes of the character, providing

[10] Nick Yee, "The Daedelus Project," http://www.nickyee.com/daedalus/

a first-person perspective. This is the perspective provided in the so-called first-person shooter games, but in a typical MMO, most players position the camera just behind and slightly above the character, thereby providing a perspective that offers a view similar to that of the character yet at the same time allows the player to observe the character at all times. This is usually the view provided to the beginning player by default, and it seems to offer various pedagogic benefits, such as a greater willingness to take risks (the character is "me," yet I am partially removed from the action) and a perspective that encourages reflection on one's actions and their consequences. This perspective provides an obvious—and established—benefit for educational purposes.

A lot of game play involves the completion of "quests," or tasks, given upon request by certain NPCs. Completion of a quest, together with the successful performance of various other actions, results in the player gaining various rewards, such as game money, equipment for the character, or "experience points." The latter, in particular, constitute a significant feature of the sense of accomplishment in playing such a game, and provide a constant measure of advancement. With greater experience, a character acquires more abilities and is able to journey to more of the game world and carry out a greater variety of tasks. Experience points are grouped in bands that constitute the levels mentioned above.

The much-prized process of "leveling up" can be viewed as a game analog of, say, high-school or college graduation. Game developers spend a huge amount of time and effort designing the leveling system in a game, and constantly adjust it to ensure the maximum effect. For the developer of an entertainment game (such as *World of Warcraft*), that desired effect is that the player wants to keep playing. As I noted earlier, one of the major attractions of video games from an educational perspective is that the game reward system can be used to motivate and encourage players to learn. This is why this is a book about the use of video games—and not just their immersive virtual world environments—in mathematics education.

Here is a preview of the four components of my argument in favor of using video games in virtual worlds for math education:

1. the immersive environment of a video game is an ideal one in which to learn everyday mathematics (and which can be designed to provide many examples of everyday math),
2. the game can provide structure to the learning; for example, by guiding the play along a dimensions-of-variation path, as advocated by requirement number 5 in my list of eleven desirable features of an ideal mathematics learning environment (see page 28),
3. the game can provide the incentive for the player to keep playing—and in so doing to keep learning,
4. both the environment and the game can be pleasurable and stimulating, two important prerequisites for good learning.

This summarizes why I believe games in virtual worlds offer the best kinds of video games for mathematics learning. MMOs are the cream of the crop, but single-player games in virtual worlds also offer potential. I have already mentioned the math ed first-person shooters *Timez Attack* and *DimensionM*. Another excellent example is game studio Valve's highly acclaimed *Portal*, a first-person shooter which, while not an explicitly math ed video game, nevertheless has significant potential to develop mathematical thinking.

Those three examples all have full 3D virtual worlds. You lose some sense of immersion when you drop the 3D effect, but there is still scope for a great educational game. Two excellent games that have 2.5D worlds are Sid Meier's strategy games *Pirates* and *Civilization*. Neither was explicitly designed as an education game, and both were hugely popular entertainment games, but they are clearly educational in their effect on players.

At an even simpler level of graphics, two good mathematics education video games having 2D virtual worlds are Mind Research's *JiJi* game and Dreambox Learning's math game for K-3 children.

When you omit the virtual world altogether, you definitely lose a significant component. Even without it, however, there remain enough features to provide powerful and effective mathematics learning. It would be unwise to restrict attention to just one genre. Certainly, the current crop of math ed video games on the market barely scratches the surface of what the medium offers.

Euclid Would Have Taught Math This Way

Observation A: Schoolchildren in the United States (and several Western European countries) are consistently outperformed in international comparison tests of mathematical ability. Teachers complain that many students appear uninterested in the subject and are unmotivated to make the effort necessary to progress in developing computational skills, problem solving ability, or an understanding of basic mathematical concepts. "They simply do not seem willing to put in the effort to learn some skills that could be of real use in their adult lives," is an often-heard remark.

Observation B: The vast majority of schoolchildren in the United States and those same countries—97% according to the Pew survey—spend many hours each week playing video games. (So too do over 50% of adults, according to Pew, but my present focus is on schoolchildren.) During the course of that game play, they may acquire a vast amount of knowledge about the imaginary world portrayed in the game, they will often practice a skill many times until they are fluent in it, and they will perform a particular action (such as manufacturing an artifact or killing a particular kind of beast) repeatedly in order to complete a quest and thereby advance in the game.

If you were put in charge of reorganizing education to solve the problem highlighted in Observation A, what would you do? If you don't immediately connect Observations A and B, and think, "Use video game technology to teach basic school mathematics," then you probably have not played any good video games, and you almost certainly have a wildly inaccurate conception of what the best and most popular of those games involve.

In this book, I am trying to convince you that the solution to the problem of unsuccessful and unmotivated students of mathematics is to make extensive—and yes, I mean extensive—use of video games in mathematics education. In fact, I am going to claim much more than that. Video games are not just a way of resolving a current crisis in mathematics education. Nor does their use in mathematics education amount to "selling the subject short," "dumbing down the material," or "pandering to the latest technology craze." Rather, for a whole range of reasons I enumerated in Chapter 2, immersive virtual worlds are in fact the ideal medium for developing basic mathematical skills and understanding, and video games are an ideal way to make use of those environments to help students learn mathematics.

Just as the invention of the printing press in the fifteenth century made it possible for everyone to acquire basic literacy, so too future generations will recognize the development of video game technology in the early twenty-first century as the time when a technology appeared that enabled everyone to acquire basic mathematical ability. Recall that by "everyday math" or "everyday mathematical ability" I mean, in general, the mathematics that is taught—or *should* be and is *supposed* to be taught—in the elementary and middle grades of schools to children 5 through 14 years old. This includes basic number concepts, arithmetic, arithmetical relationships, multiplicative and proportional reasoning, numerical estimation, elementary plane geometry, basic coordinate geometry, elementary school algebra, quantitative reasoning, basic probability and statistical thinking, logical thinking, algorithm use, problem formation (modeling), problem solving, and sound calculator use. (For reasons I will make clear later, I believe that video game technology is completely unsuited to teaching more advanced mathematics—though it can play a significant role in education at that level. My present focus is exclusively on everyday math.)

I would note that if you have managed to get along in life well enough without mastery of this everyday math, then that may simply reflect other strengths you have, but in the future faced by today's children, everyday math is going to be even more important for good citizenry than it is today.

A New Approach

I have already emphasized that it's not enough to read about video games. They must be experienced. That's why I keep urging you to go out and get a good video game (e.g., my favorite game, *World of Warcraft*) and play it. One of the most difficult lessons to learn for those of us who were successful in learning primarily through books and lectures is that games are all about experience; they can be understood only by playing them. Good video games present the player with a considerable intellectual challenge. In the true sense of the word, they are all about learning. It's just that the learning process involved in playing a good video game does not resemble what we traditionally think of as "education." Using games to teach mathematics is not about combining longstanding pedagogy of mathematics education with game technology; it is about using games to teach mathematics in an entirely different way. Admittedly, not all parts of elementary mathematics lend themselves naturally to learning in a game environment, and more advanced aspects of the subject definitely do not. But most of the mathematics typically taught in grades one through eight can, I believe, be learned faster and better with the aid of a properly constructed game than by more traditional methods.

I claim that using video games is the way Euclid would have taught basic mathematics had that technology been around in ancient Greece. This is not to say he would not have written *Elements*. For the benefit of my non-gamer readers, I should observe that many video games incorporate lots of on-screen text that the player must read to progress, and rapid advancement in many games requires the player to read pages of printed supplementary material as well. I suspect that Euclid's video game would have been no exception. Good video games do not replace reading; they supplement it, and often depend on it. And they don't replace the teacher either. Rather, they add a fourth leg to the existing educational support stool of teacher, textbook, and family and friends.

My remarks above were directed primarily at teachers and parents, my principal readers. But in writing this book I have a second audience in mind: the gamers themselves, for whom this book might act as a hook to draw them into the concept of video games as an educational medium. If you are a school-aged gamer, then unless you have somehow managed to survive your mathematics education so far to the point where you have a fairly good understanding of what mathematics really is, you will not be able to appreciate my claim that video games are the perfect medium to teach mathematics. "Oh no, they're going to make me play a video game where I have to keep stopping to solve math problems," you probably think. "What a drag." Well, actually, I *am* going to propose that you play games that involve solving math problems, but that doesn't mean anything like what you think it does, and you certainly won't have to stop playing the game to do the math. This is where the kinds of game I have been investigating differ dramatically from the rather dismal current crop of math ed video games on the market.

The fact is, no matter how many hours you've spent playing games—and even if you've had some experience with a video game that claims to teach math— you have not experienced a game that has been constructed to teach mathematics the way mathematics should be taught. You won't have done that for the simple reason that there are as yet no such games. Developing them, in particular, getting them right, is going to take some time. The few "math games" available today come nowhere close to achieving the potential that the medium offers. Once good math games have been developed, however, they will change fundamentally the popular conception of what mathematics is. For my gamer audience, then, I shall in the pages that follow try to explain the nature of mathematics in a way that I hope makes clear why basic mathematics and video games were made for each other. Conversely, it will become clear how and why much of the difficulty many people have learning mathematics stems not from a lack of innate ability, but from the unsuited and therefore grossly inadequate means we have historically had at our disposal to teach mathematics (see Chapter 3).

Directing my attention back to mathematics teachers, let me come right out and say it as I see it. In my opinion, the time is not very far away when anyone in the business of teaching mathematics at school level will have a professional duty to become familiar with video games. Not to do so will soon be akin to trying to teach English without being able to read. Besides explaining why this is the case, I will also spend a lot of time in this book showing you how you will be able to use video games to help students learn mathematics. True, we are not there yet; the current generation of games is still focusing on the low hanging, basic skills fruit. But that will change, and when that change finally comes it will be significant. Admittedly, I'm taking a positive view that it will happen. I do, however, see a danger that with so many skills-focused games now being produced, there may be a backlash against any use of video games when the current crop of games is found to make little difference to the development of real mathematical thinking, as will surely happen.

Finally, a third audience I have in mind for this book comprises professional game developers who want to try their hand at a mathematics-related video game. Although I am myself not a game developer, I have learned a lot about the game development process over the past seven years, and have talked at length with some very good game designers. As a result, I have a perspective to provide you, the game developer, with knowledge and insights into mathematics and mathematics education to help you build better math-related games, should you decide to venture into that territory. There is good reason to try. The field is wide open for good games that go beyond basic skills and develop mathematical thinking. Just be prepared for a long and challenging development process. The rewards will be sweet, but there is probably no shortcut to get there.

If game development tools continue to improve at their present rate, it won't be too long before all of the audiences I just mentioned start to overlap, with teachers, students, and parents all designing, if not complete games, then mathematical activities to embed in off-the-shelf educational game platforms. The same has already happened with video production, of course, with many teachers and students now creating and editing videos, either with a camera or a drawing tool.

For this reason, some of my discussions will be directed at a reader who wants either to build a math ed video game or to design a mathematical activity to embed in a game someone else is developing. For readers who have no such intention, those discussions will amount to observations about curriculum and pedagogy. In particular, I believe, along with Gee, that we teachers have a lot to learn from what makes a good video game work. Whatever perspective and experience we bring to the design and use of video games in mathematics education, a key distinction we need to understand is between factual knowledge and knowing how to do something.

Knowing How versus Knowing That

Even I have to admit that the title of this chapter is pretty audacious. Would Euclid really have taught math using video games? This is *Euclid* for heavens sake. One of the pillars of human mathematical development. Let me explain. It comes down to the questions "What is learning?" and "What is the purpose of learning?" I'll set the scene for my answer by describing two people to you.

Alistair MacIntyre has spent most of his adult life as a paramedic working for a large oil company in Saudi Arabia. In the remote desert region where he is based, he provides the only regular medical service for thousands of square miles, and routinely finds himself having to diagnose illnesses and to perform emergency surgery, as a result of which he has become highly skilled at both.

Roger Palmer is an insurance assessor, but from early childhood he has always wanted to be a doctor. In his spare time, he reads every medical textbook he can lay his hands on. Blessed with an excellent memory, by the time he was in his mid-thirties his medical knowledge was prodigious. Of course, since he does not possess a medical qualification, he has never been able to practice.

Suppose you find yourself stranded on a remote island and you become ill or need surgery. If chance were to bring one of these two individuals to your aid, which would you prefer? (Both characters are fictitious by the way, though my younger brother fits the description of Alistair.) I am pretty sure that, like me, you would much prefer it to be Alistair. You would put a much higher value on years of experiential learning than on factual book learning. Alistair might not know all the medical terminology, but over the years of actually practicing medicine, he has developed great skill as a medical practitioner. He could recognize various symptoms and know what to do. Roger's knowledge, on the other hand, is all theoretical.

Of course, it would be even better to have both present. To have Alistair treat you, but have him talk over your symptoms with Roger. And in practice, this is more or less what we get when we visit a real medical practitioner. To become qualified to practice medicine, an individual first has to take many courses and pass a barrage of examinations, gaining Roger-type *knowledge*, and then has to spend several years as an intern, under supervision, acquiring *experience* on the job, like Alistair.

The purpose of those initial years of classroom learning that medical students undergo is to accelerate, dramatically, the length of time it takes to reach an adequate level of ability to practice medicine. Without that concentrated burst of theoretical knowledge acquisition in the beginning, it would take many years of practice, working alongside an expert, to achieve a similar level of ability. That's why we do not train doctors by an apprenticeship system, like the craftsmen of old. It would take too long. But notice that the acquisition of theoretical knowledge here is a means to an end, not the end itself. Knowing medical facts is certainly important for a doctor, but in the end what we look for when we seek medical assistance is that the doctor does something for us. Ultimately, what is important is knowing *how to do things*, not merely knowing facts.

And the same is true for any form of education. The purpose of learning a foreign language is not to accumulate a body of facts about that language, but to be able to understand it, to read it, and to speak it, in order to communicate with others or to live in a foreign country. The purpose of studying architecture is to design buildings or other spaces for living. The purpose of learning physics is to be able to do physics. And the purpose in learning mathematics is to be able to do mathematics. Mathematics is primarily about doing, not knowing. This is not to say that a person who studies architectural design or physics or mathematics has to go on to be a professional architect, physicist, or mathematician. The abilities acquired from that learning can be used in many different professions and circumstances, sometimes indirectly. But even if the only tangible benefit from taking a course in architecture is that you become more aware of the spaces you live and work in, that is of value to you, You are a different person after you complete the course than you were beforehand. Taking the course has transformed you, and you see life differently. But the transformation is a result not of your having learned some new facts, but because you have learned to look at the world differently and to think a different way. That is what education is about.

I am belaboring this point because many people seem to lose sight of the purpose of education, seeing it as a process where you spend time acquiring facts, at the end of which you take a test to see how many of those facts you have retained. I call this the "filling a bucket" fallacy, which sees education as a process of acquiring a certain amount of educational content. Five bucketfuls this term, five next term, six the next, and so on. Give the student a test to make sure that the bucket is reasonably full after each course, and send him or her on to the next.

Does this sound familiar? It should. It's what most of our current formal school educational system looks like. It's hardly surprising, then, that many people think that is what education is all about. Why have we allowed it to get like that?

Well, for all its faults, the current system does have various things going for it:

1. it can be packaged and offered in a standardized, neatly organized fashion to many students, year after year;
2. it is relatively easy to measure the performance of both of the instructors and the students (as bucket fillers and buckets, respectively)—to "maintain standards" as the politicians keep insisting;
3. it does—or at least can, if done well—provide an efficient means of acquiring and measuring factual knowledge and of acquiring and measuring certain mechanical skills (i.e., skills not requiring a lot of intellectual effort to carry out).

All of these are significant factors, both from the perspective of society and educationally. The common mistake is to think that item 3 is the goal; that education is about the acquisition of factual knowledge and mechanical skills. It is not.

I'll come to the question of mechanical skills in due course. For now, I want to focus on the acquisition of knowledge as a possible outcome of learning, either in mathematics or in any other discipline. The acquisition of factual knowledge—of knowing *that*—is merely a way of speeding up and making more universally available the all-important, life-transformative, knowing *how*. Whether the acquired knowing *how* is related to a career or a profession is irrelevant as far as the purpose of learning is concerned. I am not just thinking about vocational training here. Rather my focus is education and learning in the most general sense. It could be learning to paint, play a musical instrument, be a television talk show host, be an actor, to have greater appreciation of the natural world, or even just to feel more confident that you could do something if you had to—whatever that may be.

Human life is a sequence of doings—of actions performed. As a species, we *Homo sapiens* have evolved various capacities and strategies that ensure our survival, the most characteristic of which being our language facility and our abilities to modify our actions based on reflection about past events, to anticipate future eventualities, to plan future actions both individually and with others, and to collaborate. Each of these fundamental species survival capacities is enhanced by learning. In short, learning enables us to do certain things better. That, ultimately, is what learning is about. It is why (viewed retrospectively) evolution has resulted in our having the capacity to learn.

In terms of evolution and species survival, knowing facts is not in and of itself important. Knowing facts is significant only insofar as the knowledge of those facts enables us to do certain things better. After all, books contain many facts, but a book cannot do anything. Humans can know facts and can do things. We can know something and we can know how to do something. Of the two, knowing how is what really matters to us.

In passing, I note that the aristocratic philosophers of ancient Greece elevated knowledge of facts to an intellectual status symbol. Regrettably, in certain halls of academia and in some sections of society, vestiges of that ancient culture still survive to this day, with people taking pride in seeking and possessing useless factual knowledge for its own sake. But apart from the entertainment value that factual knowledge may offer, knowing *how* is what really counts. And the Greek philosophers did have a purpose in their pursuit of theoretical, factual, and loudly proclaimed as "useless," knowledge. It was to show off to their philosopher colleagues in philosophical debate, and to demonstrate to anyone who cared— themselves for the most part—that they were different from the ordinary folk and able to pursue knowledge purely for its own sake. This is not meant as a criticism. An advanced society gains considerable value from having such individuals around.

To get back on track, knowing *that* is important only because it can help us to know *how*. A society that focused entirely on knowing *that* and paid no attention to the development of knowing *how* would quickly deteriorate. The ancient Greek society lasted as long as it did because it had a large artisan class that knew how to do things. Consequently, the development of knowing *how* should be the focus of education. I'll stress again that this does not mean that the acquisition of factual knowledge should not be part of learning. Sometimes you need sufficient factual knowledge in order to know how to do something. Moreover, knowing facts both is an efficient shortcut to knowing *how* and can be of great help in extending your knowing *how* abilities.

When we sharpen the focus from learning in general to learning in mathematics, the distinction between knowing *that* and knowing *how* becomes particularly dramatic. Mathematics is almost entirely about *doing*; in terms of factual content, the discipline hardly measures on the scale. Euclid knew that. If he had had access to a more efficient medium for teaching students how to *do* mathematics than the textbook, he would surely have used it. *Elements* would then have been at most a supplement to a video game. Hence the title of this chapter.

3 + 2 What Is "Doing Mathematics" Anyway?

It might seem a little odd to raise this question five chapters into a book on mathematics education. After all, doesn't everyone know what doing math is? Well, people generally think they know, but when you ask them, as I have on many occasions, you find that many don't really know, and of those that give a definite answer, those answers often differ from one another and in many cases differ from the answer typically given by professional mathematicians like me. So before I get into more specific details of how video games can be used to enhance mathematics education, we need to make sure we are all on the same page about what we are trying to achieve.

How Much Math Is There to Learn?

I have already made the case that mathematics education should be primarily geared toward *doing*, as opposed to knowing—that it's about acquiring knowing *how* rather than knowing *that*. The amount of material to know is extremely small. Take any term-long mathematics course, from elementary school through the end of high school, and you can write on a single postcard all of the key facts covered. In fact, the same is true right up through the end of the freshman year at a university. I know that from first-hand experience.

I was a high-school student and then a university mathematics major during the 1960s. In those days, every course culminated in a major performance exam, and those examinations had to be taken without any reference materials, which meant that the examinee had to know by heart all of the definitions, facts, and formulas. A typical exam question back then began by asking the student to write down a particular definition, formula, or theorem, and then apply it to solve a given problem. By the time I was in the upper grades of high school, I had decided I wanted to become a professional mathematician, so I was motivated to do well in the math exams. One of the techniques I adopted in order to ace the next exam was to write all of the basic definitions and formulas on a postcard and, in the weeks leading up to the exam, study them to the point of memorization on the bus to and from school each day. And you know what, there was never a course for which I required more than one postcard. (All right, I admit I did have to write fairly small on one or two occasions.) Incidentally, by the time I was at university, you could purchase professionally printed "revision cards," but I always preferred to create my own. In fact, there is a good educational argument for doing so. The act of constructing the cards, involving the choice of what to put in and what to leave out, leads to deeper understanding of the material.

But mathematics is not really about facts. It's a way of thinking. You don't measure an individual's mathematical achievement by how much mathematics he or she knows, you measure how much mathematics he or she can do. This means that mathematics, at the school level at least, is much more akin to sports than to classroom subjects such as history, geography, literature studies, or science, where simply memorizing facts can correlate with success. This is not to condone the fact that such subjects are all too often taught that way. But, as all students quickly discover, the reality is that you can get by pretty well in those subjects simply by rote memorization, even in science, while the same is not true for mathematics.

So what kind of thinking constitutes "doing math" as opposed to other modes of thinking, such as doing physics or writing a novel? It's actually quite hard—perhaps even impossible—to write down a good definition, one that captures all of mathematical thinking yet does not also apply to other disciplines. In my writings elsewhere, I have advocated the definition that mathematics is the "science of patterns." That's a good description as far as it goes, but it takes a considerable amount of explanation to make it at all precise. What kinds of patterns? What makes it a science? How is the science done? For the purposes of this book, however, it's fairly easy to say what "doing math" amounts to, since we are focusing on the mathematics typically taught in elementary and middle schools. As I noted in Chapter 3, that mathematics includes basic number concepts, arithmetic, proportional reasoning, numerical estimation, elementary plane geometry, basic coordinate geometry, elementary school algebra, quantitative reasoning, basic probability and statistical thinking, logical thinking, algorithm use, problem formation (modeling), problem solving, and sound calculator use. Notice that, with the exception of the first item (basic number concepts), everything in the above list is something that we *do*. That means that as you read this book, you should think of mathematics not as things to know, but as things to do.

It's Not All Natural

It has to be said that mathematical thinking is not something the human brain finds natural. (Though it is done using natural powers—what else is there?) It is a highly specialized and highly precise form of thinking that runs counter to millions of years of evolutionary development. The human brain evolved through natural selection to rapidly assess and cope with new and changing situations, to anticipate and plan for future events—often based on limited information—and to make judgments about other humans, also on the basis of limited information and generally with great rapidity. Our brain is a superb device for recognizing patterns (e.g., visual patterns, aural patterns, patterns of behavior), for making generalizations, for seeing similarities and connections between different situations, for qualitatively assessing threats and opportunities, and for making non-numerical comparative decisions.

Then, a few thousand years ago, our ancestors developed another way of thinking: what we now call *mathematical thinking*. Numbers, arguably the first stage of this development, were invented in Sumeria about 10,000 years ago. Geometry and trigonometry go back 2,500 to 3,000 years. Calculus is only 350 years old. More than half the mathematics that is known today was developed since the start of the twentieth century.

A lot of mathematical thinking is logical and precise (although contrary to popular belief not all of it). You need precise, logical reasoning to check and justify your answer. But how you get that answer is up to you. The current popular metaphor for the brain as a digital computer is very misleading. The brain's strengths are pattern recognition, associative memory, and approximate reasoning from incomplete information, not linear, logical reasoning. Computers reason logically; humans think analogically. Precise, logical thinking is so counter to the way our brain naturally operates that the surprising thing is not that many people find mathematics hard, but that any of us can do it at all! How is it possible?

The answer comes in two parts. First, humans have in fact been "thinking mathematically" in some sense for a lot longer than 2,500 or so years. It's just that, because the early form of "mathematical thinking" was not precise and did not involve manipulating numbers or symbols, we generally don't think of it as mathematics. But the introduction of *symbolic mathematics* (including numbers) simply identified, highlighted, and made more precise and accurate certain mental operations that humans had been doing for tens of thousands of years. For example, presumably even primitive man was able to judge which of two collections was larger, and to extend that ability to judge the larger of two collections when each is itself presented as two separate sub-collections. Adding positive whole numbers is just a more precise extension of that mental activity. Similarly, any creature that moves about on the surface of the Earth has to have spatial reasoning ability, and when you do so using precise symbolic representations, you get geometry. Once introduced, symbolic mathematics took on a life of its own and grew into a broad and rich discipline in its own right, much of which has little or no counterpart in everyday thinking. Incidentally, I am not claiming that it is a small incremental step from these basic mental capacities—perhaps we should call them pre-mathematics—to their symbolic, abstract mathematical equivalents. Indeed, the magnitude of the leap is indicated by the long time it took to take that step and the enormous difficulty many people today have mastering even elementary symbolic mathematics.

The second thing that enables us to perform mathematical thinking is our language capacity. Although our brains come equipped with certain innate mathematical notions, for the most part it is through language that we create mathematical concepts and make use of them to perform even rudimentary mathematical thinking. And it is through our language capacity that we are able to train our minds to perform the kind of precise, logical reasoning that we generally think of as characteristic of mathematics. I use the term "language capacity" rather than "language," because there are cases of individuals who, through birth defect or injury or illness, lose the ability to use language yet still are able to perform arithmetical calculations. For the most part, it is our capacity for language that provides us with the ability to do mathematics, even when that capacity is not realized in language itself. I discuss this at length in my earlier book *The Math Gene*.[1]

There is considerable experimental evidence to indicate the crucial role played by our language capacity in forming mathematical concepts and doing symbolic mathematics, but let me describe just one experiment here. This one was performed in 1999 at the Massachusetts Institute of Technology (MIT) by the French cognitive scientist Stanislas Dehaene and his colleagues.[2] The researchers assembled a group of English–Russian bilinguals and taught them some new two-digit addition facts in one of the two languages. The subjects were then tested in one of the languages. For questions that required an exact answer, when both the instruction and the question were in the same language, subjects answered in 2.5 to 4.5 seconds, but they took a full second longer when the languages were different. The experimenters concluded that numbers are closely tied to their linguistic representations and are accessed through those representations, and that the subjects in the experiment used the extra second to translate the question into the language in which the facts had been learned.

The researchers also monitored the subjects' brain activity throughout the testing process. When the subjects were answering questions that asked for approximate answers, the greatest brain activity was in the two parietal lobes—

[1] Basic Books, 2000.
[2] See Stanislaus Dehaene, *The Number Sense: How the Mind Creates Mathematics*, Oxford University Press, 1997, or my own *The Math Gene*.

the regions that house our general number sense (i.e., our ability to judge and compare sizes of collections without counting) and support spatial reasoning. When the question asked for an approximate answer, the language of questioning did not affect the response time. Questions requiring an exact answer, however, elicited far more activity in the frontal lobe, where speech is controlled. Of course, so much of mathematics explicitly involves writing things down—numerals, formulas, definitions, equations, diagrams, theorems, proofs—that mathematics is clearly in large part a symbolic/linguistic activity. What the results of researchers like Dehaene and others show is that language is crucial even in those parts of mathematics that are not overtly linguistic.

It is important not to lose track of the fact that although almost all of mathematics is abstract, logical, and precise—things the human brain is not naturally good at—it does come from the everyday world of human experience. This is true even for the far reaches of highly abstract, advanced mathematics, though the path that leads from the everyday world to the more advanced parts of mathematics sometimes involves several steps. The more familiar parts of mathematics all arose directly from attempts of our ancestors to meet certain changing needs of society. As far as we know, numbers were invented in Sumeria around 10,000 years ago, when the growing sophistication of their society created a need for an efficient means of mediating trade and keeping accurate account of an individual's worth (i.e., money). Various ancient civilizations invented geometry and trigonometry to keep a record of the motion of the sun, the moon, and the stars to predict crop cycles, for occult purposes, to determine land boundaries, for navigation, and to design and construct buildings. Calculus was invented in the seventeenth century to make accurate predictions about objects that move or change continuously, such as the positions of the sun and the planets in the sky. Probability theory was invented around the same time to make accurate predictions about the likely outcomes in games of chance.

When you look at the way mathematical thinking developed to meet the everyday need for more precise reasoning, you can begin to appreciate the frequently heard remark that mathematical thinking is "just formalized common sense." While that phrase does have a large grain of truth in it, it is, however, misleading in that it requires considerable mental effort to train the mind to "formalize" common sense in the appropriate way. And while defensible for everyday math, the view of mathematics as "formalized common sense" is just plain wrong for parts of advanced mathematics, where methods are often highly contrived and results can be counter-intuitive.

Math Is Simple. Honest.

The essence of thinking mathematically about the world is simplification. When we count the items in a collection, we ignore all the particulars of the things being counted—each item is simply a unit in the count. When we measure the length of some object, it doesn't matter what the object is made of, what color it is, what its temperature is, or what it is used for. We simply determine a single number—the length. When we measure the area of a floor, we know it is irrelevant to the number we get what material the floor is constructed from, its texture, its color, or its temperature. When we do plane geometry or trigonometry, we "reduce" the world to flat surfaces, straight lines, rectangles, circles, cubes, spheres, and numerical angles. It is because doing mathematics ignores almost all of the complexity of the everyday world that the application of precise, mathematical thinking yields accurate answers. The mathematical approach to the world involves a trade off of complexity for precision.

If you choose to be particularly pedantic, mathematical reasoning is not about the real world; rather, it is about a highly artificial, simplified abstraction of the real world. To take just one example, there are no perfect circles in the world, just shapes that are approximately circular. Nevertheless, the mathematical conception of a perfect circle has proved to be extremely useful. More generally, the reason mathematicians have continued to develop the subject for over 3000 years—and society has seen fit to support them in that activity—is that it has turned out to be extremely useful across the board. Mathematicians may spend their time thinking about idealized abstractions that do not, strictly speaking, exist, but people can use the results of mathematical calculations to do real things in the world; things that we would otherwise not be able to do.

Mathematics is so useful that even when there is no inherent numerical abstraction in some real world phenomenon, we invent one. Temperature is a good example. Things in the world have different levels of molecular activity that manifest themselves in some things being warmer or colder than others or warmer or colder than they themselves were an hour earlier. But this is not a numerical phenomenon. Nevertheless, it is so useful to have numbers that scientists invented a concept known as temperature, whereby a number is associated with the degree of warmth in a body. In fact, a lot of modern physics is like this!

Part of learning mathematics, of learning to do mathematics, or thinking mathematically, is becoming familiar with the particular view of the world that mathematics requires, and acquiring the ability to strip away the complexity that is present in the world and thereby reduce the world to the pure abstractions of mathematics. If you are unable to make this step—even more so if you do not realize it is necessary—then mathematics will seem to be a highly contrived, artificial symbol game having little to do with the real world we live in. And almost certainly, it won't make sense. To put it another way, doing math means thinking like a mathematician.

This, I think, hits the nail right on the head regarding one of the major problems with current school mathematics education. Hardly any math teachers are mathematicians, and a great many have not studied mathematics to a sufficient degree to know what it means to think like a mathematician. You can't blame the individual teacher, doing the best he or she can, but with the best intentions in the world you can't teach someone to think a certain way if you don't know what that way of thinking is. No one would hire a music teacher who is not a musician or an art teacher who does not enjoy and somehow practice or pursue art. Why does the same not apply to mathematics? My reason for relegating this important point to an observation well into the book is that my purpose is not to focus on the problems in current mathematics education. They are, at least for now, a fact of life. I want to spend my time looking at what can be done to make up for those failings, both by providing supplementary learning resources to the existing educational system, and by providing learning opportunities outside the system.

What Does (or Should) Learning Math Mean?

If doing math means thinking like a mathematician, then learning math must mean learning to think like a mathematician. While the logical step to this conclusion is pretty obvious, in my experience very few students in math classes ever realize this fact. In large part it is because their teachers don't realize it either—usually because they are not themselves mathematicians. Thinking like a mathematician means more than being able to do arithmetic or solve algebra problems. In fact, it is possible to think like a mathematician and do fairly poorly when it comes to balancing your checkbook. (For example, I have never been particularly good at arithmetic. I suspect that I could become good at it if I ever had reason to do it regularly, as the study of the Recife street traders suggests, but in my life I've never had to put it to the test, so we may never know.) Mathematical thinking is a whole way of looking at things, of stripping them down to their numerical, structural, or logical essentials, and of analyzing the underlying patterns. Like most people, when I am doing something routine, I rarely reflect on my actions. But if I'm doing mathematics and I step back for a moment and think about it, I see myself as a *mathematician*.

"Well, duh!" I hear you saying. "You *are* a mathematician." By which I assume you mean that I have credentials in the field and am paid to do math. But I have a similar feeling when I am riding my bicycle. I'm a fairly serious cyclist. I wear skintight Lycra clothing and ride a $4,000, ultralight, carbon fiber, racing-type bike with drop handlebars, skinny tires, and a saddle that resembles a razor blade. I try to ride for at least an hour at a time four or five days a week, and on weekends I often take part in organized events in which I ride virtually nonstop for 100 miles or more. Yet I'm not a professional cyclist, and I would have trouble keeping up with the Tour de France racers even during their early morning warm-up while they are riding along with a newspaper in one hand and a latte in the other. But in addition to biking regularly, I read cycling magazines, I follow cycling news, and I watch the major races on TV. I have even ridden up some of the classic Tour De France mountains. Being a bike rider is part of who I am. When I am out on my bike, I feel like a cyclist. And you know, I'd be willing to bet that the feeling I have for the activity is not very different from Lance Armstrong's when he was winning all those Tour de France competitions.

It's very different for me when it comes to, say, tennis. I do not follow the sport closely at all, I don't have the proper gear, and I have never played enough to become even competent. When I do pick up a (borrowed) racket and play, as I do from time to time, it always feels like I'm just dabbling. I never feel like a tennis player. I feel like an outsider who is just sticking his toe in the tennis waters. I do not know what it feels like to be a real tennis player. As a consequence of these two very different mental attitudes, I have become a pretty good cyclist, as average-Joe cyclists go, but I am terrible at tennis. The same is true for anyone and pretty much any human activity. Unless you get inside the activity and identify with it, you are not going to be good at it. If you want to be good at activity X, you have to start to see yourself as an X-er—to act like an X-er.

Semiotic Domains: Being an X-er

A large part of becoming an X-er is joining a community of other X-ers. This often involves joining up with other X-ers, but it does not need to. It's more an attitude of mind than anything else, though most of us find that it's a lot easier when we team up with others. The centuries-old method of learning a craft or trade by a process of apprenticeship was based on this idea. Gee[3] uses the term *semiotic domain* to refer to the culture and way of thinking that goes with a particular practice—a term that reflects the important role that language or symbols plays in these "communities of practice," to use another popular term from the social science literature. Notice that mathematics, cycling, tennis, movies, and video games, for example, each develop their own specialized language, generate specialized literature, and even produce insider humor that members of the specialized community may find hilarious but which leaves outsiders baffled.

In Gee's terms, learning to X competently means becoming part of the semiotic domain associated with X. Moreover, if you don't become part of that semiotic domain you won't achieve competency in X. Notice that I'm not talking here about becoming an expert, and neither is Gee. In some domains, it may be that few people are born with the natural talent to become world class. Rather, the point we are both making is that a crucial part of becoming *competent* at some activity is to enter the semiotic domain of that activity. This is why we have schools and universities, and this is why distance education will never replace spending

---

[3] James Paul Gee, *What Video Games Have to Teach Us About Learning and Literacy*, p. 18.

a period of months or years in a social community of experts and other learners. Schools and universities are environments in which people can learn to become X-ers for various X activities—and a large part of that is learning to think and act like an X-er and to see yourself as an X-er. They are only secondarily places where you can learn the facts of X-ing; the part you can also acquire online or learn from a book.

In the case of mathematics, it comes down to the point I began this section with: learning mathematics—learning to *do* mathematics—means learning to think and act like a mathematician and to see yourself as a mathematician. Not necessarily a good mathematician or a professional mathematician (though the community's doors are always open for new members to enter). I am not talking about what it takes to be *good* at math. I am talking about what it takes to be *competent* at it—like me on my bike.

The social aspect of learning that goes with entering a semiotic domain is often overlooked when educational issues are discussed, particularly when discussed by policy makers rather than professional teachers. Yet it is a huge factor. Humans are social creatures that need and seek membership in, and approval by the other members of, certain groups, and are strongly influenced by those communities. The community may involve regular or occasional face-to-face contact, but it does not need to. It may have specific criteria for membership or it may be quite loose and organic. The key is that the individual sees him or herself as a member of the community and is acknowledged as such by other members. For example, people who pursue sporting activities, sports fans, fans of specific television series, members of clubs, members of political parties, and professionals in various domains all manifest the characteristics of this deep-seated human trait.

Video games also generate their particular communities of practice, with committed players coming to view themselves, and act as, members of that community. This is true for single-person games as well as for MMOs, though with the latter the community aspect is an explicit and integral part of the game play. Experienced educators know the enormous educational power of the social aspect of learning and try to make use of it in the classroom. Video game developers recognize the key role that social activities associated with, and perhaps part of, a game can play in its success. A video game that players want to play and that generates an enthusiastic community would be well on the way to achieving significant educational objectives.

Armed with this general orientation of what math ed video games should be trying to achieve, it's time to get to the details regarding the mathematical abilities we (the education community) want the players to acquire.

Mathematic Proficiency: A New Focus in Mathematics Education

By now, I hope I have managed to convince you that video games provide an excellent—I believe ideal—technology environment for learning everyday mathematics. What features should a video game have in order to provide an effective mathematics learning experience? What kinds of game develop mathematical thinking? In this chapter, I outline the basic pedagogic principles on which I think the design should be based. Those principles come with impressive credentials—the National Research Council (NRC)—and are widely accepted in the United States mathematics education community. Some of the principles—there are five of them—may seem to bear little or no relation to the mathematics classes you had in school. The blue-ribbon committee that the NRC engaged to develop the principles did its work around the turn of the new millennium, and their mission was to develop an educational framework that would prepare students for life in the twenty-first century. Much of what they recommended has yet to find its way into mainstream American mathematics education.

The Five Strands

Until the first half of the twentieth century, learning school mathematics usually meant becoming proficient in performing the procedures of arithmetic, with relatively little attention paid to students' understanding. In the 1950s and 1960s, the "New Math" movement called for the focus to be shifted to understanding the concepts of mathematics (e.g., number concept, number bases, laws of arithmetic). Although the concept was sound—focusing on conceptual understanding *and* computational skill—in transferring New Math from the universities, where the idea originated, to the nation's classrooms, where many teachers were ill equipped to handle the new material properly, the New Math idea was transformed in the minds of many to conceptual understanding alone. A popular joke at the time was "I don't care if you get the wrong answer, as long as you understand the method."

The backlash created by New Math, or rather a result of the popular conception of New Math, was the rapid emergence of what was called by its proponents the "Back to Basics" movement, where the focus was once again on developing the ability to compute rapidly and accurately. The 1980s and 1990s saw the emphasis shift yet again, this time toward what was called "mathematical power", which involved—in addition to arithmetical skill—logical reasoning, problem solving, connecting mathematical ideas, and communicating mathematical ideas to others. The mathematical power initiative led to two schools of thought, with one group once again promoting the primacy of memorization and the acquisition of rote-learned skills, and another arguing that, in addition to problem solving skills, students needed to be able to prove mathematical assertions.

Meanwhile, as the educational sands kept shifting and researchers and educators argued the different viewpoints, many parents maintained a fairly constant viewpoint. Middle-class parents with sufficient means employed private tutors or enrolled their children in commercial Saturday morning academies where the emphasis was predominantly on the development of arithmetical skills through repeated practice—much like the kind of mathematical education you would have found in the church schools of medieval Europe.

Today, there is a substantial consensus within the mathematical community centered on what has come to be called *mathematical proficiency*. This concept attempts to codify the mathematical knowledge and skills that are thought to be important in today's society. The elements of mathematical proficiency were spelled out in a report from the National Research Council's 1999–2000 Mathematics Learning Study Committee, titled *Adding it Up: Helping Children Learn Mathematics*, published by the National Academies Press in 2001. In my view, that volume provides the best single source for guidelines toward good mathematical instruction to date. I would strongly recommend that any designer of a math ed video game takes this as an initial guide.

From the outset, however, you should recognize that the NRC report, together with practically everything that has been studied and written regarding mathematics education, is based on a classroom model involving the three-way interactive triumvirate:

For successful education in this environment, the teacher has to understand both what is being taught (the mathematics) and what is involved in learning mathematics (the student's part), and the student has to interact with both the mathematics and the person teaching it. Almost all current pedagogic theory and practice is based on this model.

But this is very different from the learning framework you will be working with when you select—or build—a math ed game. No student has yet experienced mathematics learning with a video game, where the framework looks like this:

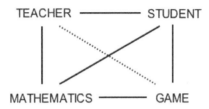

There are no experts in the use of this kind of educational framework. There is no body of accumulated knowledge that you can draw upon. While you can and should take *Adding It Up* as a starting point and use it and the existing mathematics education research literature as a constant reference source, once you start you will be largely out on your own. The best you can do is keep one primary goal uppermost in your mind. The game should help students achieve mathematical proficiency, as articulated in *Adding It Up*, and it should do so using activities that arise naturally in the game.

Mathematical proficiency is presented in the report as having five interwoven strands[1] (see Figure 4):

- *conceptual understanding*—the comprehension of mathematical concepts, operations, and relations;
- *procedural fluency*—skill in carrying out arithmetical procedures accurately, efficiently, flexibly, and appropriately;
- *strategic competence*—the ability to formulate, represent, and solve mathematical problems (arising in real-world situations);
- *adaptive reasoning*—the capacity for logical thought, reflection, explanation, and justification;
- *productive disposition*—a habitual inclination to see mathematics as sensible, useful, and worthwhile, combined with a confidence in one's own ability to master the material.

---

[1] See *Adding it Up: Helping Children Learn Mathematics*, National Research Council, National Academies Press, 2001, p. 116.

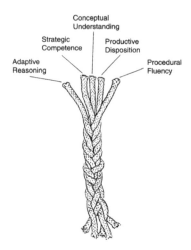

Figure 4. Intertwined strands of proficiency.[2]

The authors of the NRC report, as well as other proponents of mathematical proficiency, stress the importance of viewing all five strands as tightly interwoven. For instance, the authors state in *Adding it Up* (the National Research Council Report, or NRCR):

> The most important observation we make here, one stressed throughout this report, is that *the five strands are interwoven and interdependent in the development of proficiency in mathematics.* Mathematical proficiency is not a one-dimensional trait, and it cannot be achieved by focusing on just one or two of these strands. ... [W]e argue that helping children acquire mathematical proficiency calls for instructional programs that address all its strands. As they go from pre-kindergarten to eighth grade, all students should become increasingly proficient in mathematics.
> [p. 116, emphasis in the original]

If a designer sets out to develop a game built on the principles of mathematical proficiency, one of the first questions to arise is: which of the strands of mathematical proficiency can be addressed most effectively in the game? Ideally, of course, the game should cover all five strands. Personally, I am convinced that in the long run, as the mathematics, math ed, and game developer communities gain experience in designing and using games and game environments in math-

[2] Figure taken from *Adding it Up: Helping Children Learn Mathematics*, National Research Council, National Academies Press, 2001, p. 117. Reprinted with permission from the National Academies Press, Copyright 2001, National Academy of Sciences.

ematical education, games will eventually be produced that cover all five strands effectively. But at the moment, given our limited experience in these learning environments and typically limited resources, game developers will inevitably have to make choices—just as schools do. The question then is: where will the game have the greatest impact? What elements of mathematical education can be done better with a video game than by a teacher in a classroom without such an aide? Where can we take advantage of video game dynamics to compensate for weaknesses in other methods of instruction? A math ed video game is, after all, just one additional medium for helping an individual acquire mathematical skill. In terms of making a positive contribution to mathematics learning, coverage of the five strands should come from the entire educational environment: school, home, society, and additional media such as video games.

With those thoughts in mind, let's take a closer look at what mastery of each of the five strands of mathematical proficiency entails.

Conceptual Understanding

Conceptual understanding of a mathematical topic is more than knowledge of a collection of isolated facts and procedures. Rather it involves organization of the material into an integrated, coherent whole. A student who has achieved conceptual understanding is better able to learn new facts and methods by connecting them to what he or she already knows. Since humans best remember things that make sense to them—things they understand—conceptual understanding also supports retention. A student who understands a method—why it involves the steps it does and when it may be applied—is unlikely to remember it incorrectly or to misapply it. A student who understands numbers and arithmetic will realize that a calculation performed on a hand calculator that displays $19 \times 17 = 2{,}223$ must be wrong, since both the given numbers are less than 20, and $20 \times 20 = 400$, so most likely there has been an entry error. (In this case, pressing one key twice and erroneously entering 19×117.)

Recognizing if or when a student has achieved conceptual understanding is difficult, even for a trained teacher with many years of experience. For an automated instruction system, whether or not it is embedded in a video game, such recognition is virtually impossible, regardless of how much artificial intelligence (AI) is included. Possibly the only manifestation of conceptual understanding that is relatively easy to measure is the student's ability to transfer from one representation of a mathematical situation to another. But this is a decidedly one-directional metric. A student who did not correctly convert fractions to decimals or vice versa, or did not correctly pair fractions with geometric representations of fractional quantities such as subdivided lines or sectored circles ("cut up pies") most definitely does not yet have a conceptual understanding of rational numbers. But carrying out these tasks and getting the right answers does not in itself indicate conceptual understanding; the student may simply have memorized the rules without understanding them, or may even have guessed.

Without the close attention of a skilled teacher, it is possible for even a highly intelligent and mathematically motivated student to fail to acquire correct conceptual understanding. For example, a student who has memorized the cross-multiplying procedure for adding fractions and who produces the calculation

$$3/4 + 5/7 = (3 \times 7 + 5 \times 4)/(4 \times 7) = (21 + 20)/28 = 41/28$$

may be classified as having "understood fractions", whereas a student who produces the incorrect addition

$$3/4 + 5/7 = (3 + 5)/(4 + 7) = 8/11$$

will generally be thought to "not understand" addition of fractions. Yet the second student might have reflected on the matter and argued:

> If I have 4 red balloons of which 1 is burst, then 3/4 of my red balloons are good. If I have 7 blue balloons and 2 of them are burst, then 5/7 of my blue balloons are good. So what fraction of my balloons are good? Well, altogether I have 4 + 7 = 11 balloons, and of them 3 are burst, leaving 8 good ones, so the fraction of my balloons that are good is 8/11.

This is perfectly correct reasoning, showing that the student understands the concept of fractions. The student's problem is that he or she is unaware that, for very good reasons, the mathematical community long ago decided that adding fractions means something different from adding proportions—even though fractions may be and are used to quantify proportions. It would take a skilled teacher to identify the second student's mistake here, and help that student make the transition to the "correct" definition, without making him or her come to the conclusion that mathematics is arbitrary and illogical. Simply telling the student "Your way is wrong, let me show you the right way" is likely to be a disastrous strategy, particularly for a student who has an inclination to think things out for him or herself. The teacher has not only to show the student the "correct"—i.e., accepted—method, but also must explain to the student exactly why mathematicians decided it is more useful to define addition of fractions the way we do. I'll return later to the highly problematic issues of achieving and recognizing conceptual understanding, but for now let me just mention that the acquisition and mastery of procedural skills—our next topic—is a key step toward conceptual understanding. In particular, some educators have suggested that understanding should always precede mastery of skills, which in my view is simply wrong.

Also misguided, though perhaps not wrong, is the suggestion that understanding should at least always accompany mastery of skills, so that the student never has to experience working with concepts or procedures he or she does not understand. While perhaps doable in theory, for many mathematical topics it would take far too long, with a likely result that the student would simply lose heart and give up long before achieving sufficient understanding to start to do useful mathematical thinking. But as language creatures—the "symbolic species" to use neuroscientist Terrence Deacon's term[3]—we have a powerful ability to learn to follow symbolic rules without understanding what they mean; and as pattern recognizers with an instinct to perceive meaning in the world, once we have learned to play such a "symbol game" it does not take us long to figure out what it means. Playing chess is an excellent example of how learning to play by simply following the rules rapidly leads to an understanding of the game. Thus, while the idea that students should always "understand what they are doing" has a lot of appeal, it ignores that fact that nature has equipped us with a far more efficient method of learning. Notice that I am not saying understanding is not important. It is. Mastery of skills without understanding is short-lived and brittle. Conceptual understanding is always an important goal in mathematics education. It's what it takes to get there that I'm focusing on.

Procedural Fluency

NRCR describes procedural fluency as knowledge of procedures, knowledge of when and how to use them appropriately, and skill in performing them flexibly, accurately, and efficiently. Echoing the observation I made a moment ago regarding the role played by procedural fluency in attaining understanding, NRCR goes on to observe that "In the domain of number, procedural fluency is especially needed to support understanding of place value and the meanings of rational numbers. It also supports the analysis of similarities and differences between methods of calculating."

NRCR explicitly includes the use of calculators as one of those "methods of calculating." Given the ubiquity of cheap, efficient calculators, there is clearly little practical need for students to become fast or efficient at multi-digit, paper-and-pencil or mental arithmetic involving large numbers. However, many tasks involving mathematics in everyday life require facility with carrying out algorithms, and thus it is important that students become proficient in performing algorithmic

---

[3] Terrence Deacon, *The Symbolic Species: The Co-Evolution of Language and the Brain*, W.W. Norton, 1997.

computations. Moreover, it is important for every citizen to have a good understanding of the place-value decimal number system that forms the bedrock of all
science, technology, business, commerce—indeed pretty much all of contempo-
rary life—and it is hard to see how that could be achieved except through the mastery of the basic arithmetic algorithms for addition, subtraction, multiplication,
and division. That should include the long division algorithm, in my view, because
it makes the most significant use of the place-value system.

Returning to the theme of achieving conceptual understanding, NRCR, in its
discussion of procedural fluency, notes that

> [A] certain level of skill is required to learn many mathematical concepts
> with understanding, and using procedures can help strengthen and
> develop that understanding. For example, it is difficult for students
> to understand multi-digit calculations if they have not attained some
> reasonable level of skill in single-digit calculations. . . . Without sufficient
> procedural fluency, students have trouble deepening their understanding
> of mathematical ideas or solving mathematical problems. The attention
> they devote to working out results they should recall or compute easily
> prevents them from seeing important relationships. Students need
> well-timed practice of the skills they are learning so that they are not
> handicapped in developing the other strands of proficiency. (p. 122)

In my view (and that of others), there is only one way to achieve fluent proficiency at any skill, either physical or cognitive: *repetitive practice*. In mathematical education, this was known–or at least accepted—for many centuries, but was
questioned during the course of the past half century in large part because the
more stimulation-rich environment in which children now grow up makes them
less prepared to endure sufficient training cycles to achieve mastery of mathematical procedures. The "progressive" changes to the school mathematics curriculum
that were introduced during this period were motivated in large part by both a
desire to make the material sufficiently interesting to maintain student interest,
and to *avoid* unmotivated repetitive rote learning. True, some educators may have
gone overboard and seemed to advocate avoidance of repetitive practice altogether, but they were surely in a minority, though they still drew a lot of critical attention. Repetitive learning is not the problem. Humans of all ages show no
disinclination to practice sufficiently to acquire new skills when they have a good
reason to do so. Just watch a child learning to skateboard, to ski, to snowboard, to
play a musical instrument, or to become skilled in a video game. What makes this
learning process work is that the learner is focused on a desirable goal. Aha! This
touches on why I believe that procedural fluency is one area where video games
can have an enormous positive impact on mathematics education.

Learning basic mathematical skills is no more difficult than any of the above examples, and in fact simpler than many. For example, because of the various easily recognized patterns and other symmetries in single-digit multiplications, learning the multiplication tables comes down to rote memorization of at most about a dozen facts.

As I argued in my book *The Math Gene*,[4] apart from a very small number of individuals with a recognized cognitive impairment (*dyscalcula*) anyone is capable of acquiring procedural fluency in basic arithmetic, and the main problem facing the achievement of such fluency is not the amount or the complexity of the material but the level of abstraction. My main focus in *The Math Gene* was to understand how, in evolutionary terms, the human brain acquired the capacity for mathematical thinking, not mathematical education. Thus, I did not go into what is required for an individual to take advantage of that innate capacity to become proficient in mathematics where other factors come into play, in particular motivation and social influences.

Once it has been recognized that

- anyone can, in principle, become procedurally fluent in mathematics and
- it requires a considerable amount of repetitive practice to acquire that procedural fluency,

then learning basic mathematics comes down to

- motivation to practice,
- positive feedback,
- maintenance of interest.

This is where video games provide a huge advantage—particularly for students who have grown up in the instant-feedback, multi-tasking (INFEMUT) era of MTV, computers, social networking, instant messaging, and sophisticated, high quality video game technologies. I shall examine this issue in more detail in due course. In the meantime, let's turn to the third of the five strands of mathematical proficiency.

Strategic Competence

This is the ability to formulate, represent, and solve mathematical problems, based on some real-life, simulated real-life, or (in our case as video game designers) simulated fantasy-life, situation. The best known—and arguably the dullest and least imaginative—way to develop this ability are the familiar "word problems" (also called "story problems") of mathematics textbooks.

---

[4] Basic Books, 2000.

Representing a real-world problem mathematically involves, among other things, deciding which features of the situation should be represented and the kinds of representation (e.g., numerical, graphical, symbolic) to be used. It also involves recognizing what is known and what is to be found, recognizing relationships that are instances of properties, and perhaps expressing these so as to represent the known in terms of the unknown, leading to algebraic solution. Many word problems can be done by arithmetic, but may need deep insight into the relationships being called upon.

Solving the problem is likely to involve recognition that, when represented mathematically, problems from different kinds of situations may exhibit similarities that may be exploited in order to find a solution. For example, a solution to a problem involving fluid flow along a river may be useful in analyzing the flow of money in financial markets. (A Nobel Prize was awarded to the two people who first took advantage of that observation.) Some problems of this kind are routine, requiring little more than that the student recognizes the general type of problem and knows what the appropriate approach is. But others may require greater flexibility and originality. Practice at dealing with problems of both kinds is important in developing strategic competence.

Studies such as the Recife street market children, the Adult Math Program, and others, have shown that people perform considerably better at word problems about a real-world activity (especially one that is familiar and meaningful to them) than they do at symbolic representations of the same mathematics, but far less well than they do in the actual real world situation. It is not hard to understand why. What people become good at with practice is doing real-world math (generally arithmetic) as the need arises in a real-world situation, when they are reasoning not about symbolically represented mathematical abstractions but about real things, particularly things over which they have some control and where the consequences matter to them. A word problem will succeed in tapping into this capacity to the extent it puts the individual into the same cognitive state as the corresponding real-world situation.

A good novelist can use words to create in the mind of a reader a feeling of "being there," and perhaps in such a cognitive state the person would be able to "do math," but a glance at typical mathematics word problems makes it abundantly clear that they are a far cry from world-class literature.[5] They do not come close to creating a sense of being in a real-world situation of the kind they purport to describe. Indeed, the overpowering sense one gets when reading a word problem is simply one of being faced with having to do a word problem!

[5] The whole issue of word problems in mathematics education was examined in depth by Susan Gail Gerofsky in her 1999 doctoral dissertation at Simon Fraser University in Canada, "The Word Problem as Genre in Mathematics Education."

In contrast to word problems, however, video games (particularly those in a virtual world) are ideally suited to presenting the student—the player in this case—with stimulating challenges that require, encourage, exercise, and develop strategic competence.

Adaptive Reasoning

The phrase "mathematical reasoning" is often taken to mean step-by-step logical deductive reasoning and formal proofs. The NRCR uses the term "adaptive reasoning" to refer to something much broader, which includes "informal explanation and justification" as well as "intuitive and inductive reasoning based on pattern, analogy, and metaphor." (p. 129) Broadly speaking, this is the kind of reasoning that professional mathematicians engage in on a daily basis—as opposed to the arguments they present in published work and in mathematics textbooks, which tend to be restricted to formal proofs. A player's successful performance in many current video games, such as *World of Warcraft*, involves a considerable amount of this kind of reasoning, though rarely of an overtly mathematical nature. Thus, the creation of a game where explicit mathematical activities are required will almost inevitably, and without extensive special engineering, lead to the development of mathematical adaptive reasoning skill.

Productive Disposition

The NRCR provides the following description of productive disposition:

> [P]roductive disposition refers to the tendency to see sense in mathematics, to perceive it as both useful and worthwhile, to believe that steady effort in learning mathematics pays off, and to see oneself as an effective learner and doer of mathematics. If students are to develop conceptual understanding, procedural fluency, strategic competence, and adaptive reasoning abilities, they must believe that mathematics is understandable, not arbitrary; that with diligent effort, it can be learned and used; and that they are capable of figuring it out. Developing a productive disposition requires frequent opportunities to make sense of mathematics, to recognize the benefits of perseverance, and to experience the rewards of sense making in mathematics. (p. 131)

It is both interesting and relevant that virtually the entire description applies equally well to achieving success in many video games. In my view, with skillful design it should be possible to create a video game that engenders in the player a sense that mathematics is natural, useful, and worthwhile *within the context of the game*. The alternative, which designers should avoid, is that the mathematics comes across as essentially external to the game itself—simply a collection of hoops that the player must jump through in order to progress in the game. That, unfortunately, is a major drawback of all current math ed video games.

Even if a designer succeeds in fully integrating the mathematics into the game storyline, it is by no means automatic that the player will, by playing the game, come to view mathematics as useful and worthwhile in the real world. The same is true for traditional mathematics education. It's a manifestation of a general problem in education known as the "transfer problem." This is the tendency many students display of being unable to take what they have learned in one class or one context and apply it in another class or in a real-world situation. If a game exposes the player to uses of mathematics that bear a resemblance to real-life applications of math, then there will be at least a good chance that players will automatically come to develop a productive disposition. Other than ensuring that the mathematical tasks encountered in game play are realistic, however, at the moment I don't see anything specific a designer can do to achieve this goal.

Researchers who worked on situated cognition tried to circumvent the transfer problem but in the end simply rephrased it as "How does the 'situatedness' broaden its scope?" It is possible that learners, by being encouraged to generalize, by developing sensitivities to identifying patterns, by coming to recognize relationships, and by learning to distinguish between guessing randomly or wildly and developing a principled approach to trying available actions, will indeed experience an extension or broadening of the situatedness of their actions. Players in a video game should be continually prompted and encouraged to draw back from the actions that are successful and consider what it was that was that made them effective and thus can be used in other circumstances. Investigation of these issues will have to await the first release of the kinds of game I am talking about in this book.

The Key Features of Gaming

With the NRC's five strands to guide us, we are now in a position to start to rethink mathematics education to take advantage of the enormous potential of video game technology. I listed some of the more obvious features that video games provide for everyday mathematics education at the end of Chapter 3:

1. the two-dimensional or three-dimensional immersive environment of a video game (if it has one) is an ideal one in which to learn everyday mathematics (and which can be designed to provide many examples of everyday math),
2. the game can provide structure to the learning,
3. the game can provide the incentive for the player to keep playing—and in so to doing keep learning,
4. both the environment and the game can be pleasurable and stimulating, two important prerequisites for good learning.

Chapter 2 laid out the main thrust of the case for the value of feature 1, the game environment. Let's take a look at features 2 and 3, the game mechanics. (Feature 4 needs no elaboration.) Consider again Beck and Wade's list of features exhibited by video games and their regular players that we saw in Chapter 3:

1. failure doesn't hurt,
2. risk is part of the game,
3. feedback needs to be immediate,
4. used to being the "star,"

  5. trial and error is almost always the best plan,
  6. there's always an answer,
  7. I can figure it out,
  8. competition is fun and familiar,
  9. bosses and rules are less important, and
  10. used to group action and conflict.

I shall refer to these as the Key Features of Gaming (KFG). Anyone setting out to develop an educational video game—a more accurate description for any game that is to have a reasonable chance of success would be "a video game with an educational purpose"—can take advantage of these features to design effective experiences.

Many, perhaps most, of the KFG are the *opposite* of a traditional school mathematics teaching environment and the behaviors it encourages. Let's look briefly at each one.

Failure doesn't hurt. Many students give up math altogether because of the pain of failure. In a typical linear math teaching environment, failures pile up one on top of another until the student simply loses heart and gives up.

Risk is part of the game. Taking risks is an essential part of learning mathematics. Doing mathematics—thinking mathematically—does not come naturally, any more than do riding a bicycle, skateboarding, or playing the piano. To learn it, you have to go out on a limb and be prepared to fail. And you will fail. Repeatedly. None of us likes failure. Because of the fear of failure and its consequences, particularly the public failure in a typical classroom, many students quickly become risk averse in the math class. They are unwilling to try anything new.

Feedback needs to be immediate. Good teachers know this, but it's hard to achieve in a classroom setting. John Mighton, in his 2004 book *The Myth of Ability*,[1] gives an account of how effective immediate and positive feedback can be in teaching math, even to supposedly "mathematically weak" students. Although Mighton's methods have generated some controversy, I don't think anyone disagrees with his claims about the educational value of immediate positive feedback.

Used to being the "star." Many math classes gravitate to a situation where there are just one or two classroom stars, and those individuals often get lots of positive feedback from the teacher for occupying that position. But that comes at a cost to all of the other students. The game environment allows us to make every player

---

[1] Walker and Company, 2004.

a math star. The inner personal satisfaction gained in becoming a math star in a multiplayer video game will likely confirm the adage "success breeds success."

Trial and error is almost always the best plan. As I observed earlier, what gamers actually learn is that trial and error is much more effective when tempered by experience. Trial and error, guided by experience, is an integral part of the way professional mathematicians work. A more accurate description would be trial and improvement. But students in a typical math class quickly learn that trial and error is a disastrous strategy. Traditional mathematics instruction rewards students who learn the rules and practice applying them.

There's always an answer. Though this is not the case in many real world applications of mathematics, particularly at more advanced levels, it is true for the kinds of basic mathematical skills we are looking at in this book.

I can figure it out. This is coupled with the two previous key features. Such an attitude is essential to being able to learn mathematics. Yet, in a typical math class, most students come to the conclusion that when it comes to math, they *cannot* figure it out, and they look for a teacher or a textbook to do it for them.

Competition is fun and familiar. Mathematics classes where a gifted teacher can create a non-threatening, competitive environment usually produce excellent results. But for many students in a typical learning environment, competition simply contributes to a feeling of fear and dread at being "beaten" by others. At the other end of the spectrum, some students are unwilling to "win" at competition because they do not want to stand out in the competition and be ostracized by their peers. The use of a video game, even a multiplayer game, provides a safe haven for both kinds of student.

Bosses and rules are less important. This is the opposite of a traditional mathematics learning environment. Even when the teacher is kind, gentle, and understanding, he or she is unavoidably perceived as (and in some respects is) an all powerful, all knowing authority figure.

Used to group action and conflict. Many teachers encourage group work, often with some success. Few utilize conflict. The reasons in a real classroom setting are obvious and justifiable. Yet conflict is a powerful human motivator when it comes to tackling difficult challenges, which learning math most definitely is.

Clearly then, developing a video game in which students can successfully learn mathematics requires a rethinking of how students could learn math.

Learning by Exploration

As I noted earlier using different terminology, there are two alternative (and dia-
metrically opposed) models for learning math:

- learn the rules and practice using them, and
- explore the domain and "intuit" the rules.

In the 1950s when I first learned mathematics, the first model was the standard way. In many school classrooms, it still is. But today, many innovative teachers—often labeled "progressive," a term that is sometimes used as a derogatory
description by those who oppose such methods—try to adopt the second model.
I believe that choosing between the two approaches is actually not a simple matter. Given an ideal situation of (i) an unlimited amount of time, and (ii) one-on-
one coaching by a good mathematics teacher (someone who, besides knowing
how to teach this way, also really understands the mathematics), I am certain that
the second model is by far the best approach for learning basic mathematics. For
more advanced parts of mathematics, I believe that the first model is the only
practical initial approach. But in our real world, given the amount of mathematics
a typical student should learn in order to live a fruitful and rewarding life with
sufficient career choices, the second model is simply not practical—at least not as
the dominant model. But it is, in my view, far too valuable to be discarded. Later
in the book I shall argue at some length that good basic mathematics education
should involve both approaches. Although the two models seem to be contradic-
tory, there is nothing to prevent a teacher from going back and forth between
the two, developing skills and understanding at the same time, in tandem. Many
teachers do just that.

When it comes to designing a video game in which students (more precisely, *players*) learn mathematics in the course of playing the game, the choice between learning models is effectively made for us. According to the Key Features
of Gaming (particularly KFG5), to take advantage of the gaming environment we
really have to follow the exploration model. That means that the game environment must be structured so that the players' exploration moves them, overall,
in the direction of learning that we, as educators, want to accomplish. Along the
way, the student will—perhaps unwittingly—develop the all-important basic
skills.

In mathematics and in other subjects, learning by exploration has several
advantages and results in more powerful, longer lasting, and more usable knowledge than learning by instruction. Part of the reason is that when we explore, we
frequently make mistakes. Actually, from a learning perspective they are not really mistakes; rather they are choices that subsequent experience tells us were

not correct or not optimal. And as most of us know, lessons learned as a result of our "failures" are rarely forgotten. In order for failure to be an effective learning mechanism, the failure has to hurt sufficiently for us not to want to repeat it, but should not be so great that we lose heart and give up. Fortunately for our purposes, video game designers have perfected to a fine art the kind and degree of "hurt" that encourages players to improve their game play and yet does not induce them to simply abandon the game. If failure in an educational game leads to the student—the player—giving up, the game has failed to meet its objective in both gaming and educational terms.

Another advantage of learning by exploration is that it is simply more fun. We humans are inquisitive creatures, and few of us like to be simply told what to do. As Gee[2] and others have observed, successfully playing a good video game can involve a huge amount of learning, yet players rarely view game playing as a learning activity. While I certainly am not one to claim that learning must always be fun, the notion that education cannot be fun, which I believe is implicit in claims that video games are a waste of time, is likewise way off target.

Explorative learning also leads to *ownership* of what is learned. We appreciate, value, and take pride in what we discover for ourselves. And we are then motivated to put in the often considerable effort required to polish our discovery and make use of it. Good teachers know this and make use of it all the time. Successful video game designers know it too, though since they usually don't see what they do as "educational," they probably don't think of it this way.

One challenge in teaching mathematics, whether in the classroom or by incorporating mathematics learning experiences into a video game, is that much of mathematics involves learning how to do things "the right way" (or at least, "a right way"). While most subjects offer the learner constant opportunities for "doing it better," in mathematics, until you have the right answer, you have not solved the problem. This is true even when the method you adopted to get to that answer is novel and ingenious. And the fact is, no one likes to fail. It makes us feel bad. If it happens to us often enough, we will tell ourselves that whatever it is we are trying to do is not really important to us anyway, and we will give up and do something else. This risk is potentially greater in a video game than in a school classroom. A good teacher can always tell the student that his or her approach, while not leading to the correct answer, is nevertheless of merit and something to be proud of. But that kind of feedback is simply not possible in a video game, or indeed from any software system.

One way a designer can counter this risk in designing a math ed video game is to give failure at a mathematical task no worse game consequences than failure at any other game task, and to balance the math tasks and the non-math game

---

[2] James Paul Gee, *What Video Games Have to Teach Us About Learning and Literacy*

tasks so that when a player fails several times in a row in a math task, there is always an alternative game path that involves the solution of non-mathematical problems. If the designer were to leave it at that, though, this feature could lead to a player who finds math particularly hard—precisely the kind of player who would benefit most from practicing their math skills—to always opt for the non-math tasks. To counter this, the designer could adopt a basic design principle that completion of a math task will always yield a better reward than completing a non-math task. In such a game, a player could advance through to the end without ever completing a math task. But advancement will be faster and yield greater rewards in terms of acquiring "cool stuff" (better weapons and armor, cooler looking clothes, desirable possessions) if the player completes math tasks. Readers who do not see the force of the point I am making have surely never played a good video game.

That people hate to get things wrong is not the only problem facing the designer of a math ed video game. Another feature of mathematics is that in many areas of math, mastery is achieved only by repetitive practice. Mathematics is by no means unique in the need for practice, of course. It's true of many things, such as learning to play a musical instrument, skateboarding, creative writing, or driving a car. But for most of us, toleration of repetitive practice requires considerable motivation—the end must clearly justify the means.

For students who, for whatever reason, want to become good at math, repetitive practice is no problem. That nerdy kid in the class, whether in math or any other subject, who always seems to do better than anyone else almost certainly works very hard at it—often in secret—and does so because he or she seeks the self-satisfaction of being "top" in that domain. But for students who do not have sufficient desire to be good at math—and that is the vast majority of students—alternative motivators are required for them to put in the necessary hours of practice. Video games are particularly well suited to motivating repetition, and to making it meaningful and fun. There are few things in real life that children typically see as desirable goals that recognizably require mastery of math, but in a video game the designer can structure things so that the player always has strong game motivation to practice mathematical skills.

Critical to making this approach work is that the mathematics that the player has to perform has to arise naturally within the game. If the focus is on real-world, basic math skills, this is easily achieved. The designer simply has to make the game sufficiently like the real world. Indeed, the designer can put greater emphasis on the features of the real world relevant to the mathematics in question. For example, if the goal is to develop students' abilities to carry out proportional reasoning, the game should present them with lots of (naturally arising) activities that require such thinking.

Tools in the Game Designer's Educational Arsenal

This discussion requires a caveat. While the math problems that have to be solved in a video game should arise naturally in the game world, they do not all have to be solvable within the game—though to maintain a good game pace, most should be, at least at the lower levels. It is possible to design a game to help develop paper-and-pencil, symbolic math reasoning. Not only is this an extremely valuable ability that should be mastered by as many people as possible, it provides an opportunity for the game to lead to an improvement in test scores!

One way to do this is to include challenges that can be solved more efficiently if the player stops playing and works offline for a few moments. In fact, at higher levels of game play, a designer can throw in some math problems that would require several days of offline thinking to find the best solution. The player who takes time out to consider and solve the math problem not only advances faster or further in the game, she or he also learns the valuable lesson that it is sometimes better in the long run to stop for a while and reflect than to continually press forward in an exploratory mode. What is important from a game perspective is that the decision to engage in some offline reflection be at the player's own volition, and not something required by the game dynamics.

Fantasy video games offer another advantage to learning math that a designer can incorporate into the design of a game—what I call the "Pythagorean factor." There is a strong cultural/historical association between numbers and mysticism. The designers of any fantasy game that involves "ancient lore" can milk this association extensively to make math seem cool. Some of us already think it is!

A multiplayer game offers a number of potential educational benefits. First, having many players in the game can be leveraged to provide player motivation and community, since much of the incentive for players to tackle the more difficult math problems in the game will be to acquire cool stuff that fellow players can see, a recognized motivational feature of multiplayer games. Also, the sense of community that accompanies play in a multiplayer game can help motivate and sustain the interest of players who find the math parts particularly difficult. Another significant educational advantage of multiplayer games is to encourage group work to solve math problems—something many educators strongly advocate—and to take advantage of the fact that deeper understanding often comes from teaching others. One possible way to encourage more experienced players to help others with their math is to confer elevated status in the game for those who take on the role of "teacher." An obvious challenge in doing this is to ensure that when one player teaches another within the game, the instruction is correct. Of course, there is no possibility that the game designer can control what happens when one player tries to help another outside the game's own communication channels. But that is true for any form of educational environment. No teacher can prevent Sally giving Johnny incorrect advice when they get together after school to do their math homework.

Mathematics Education and Gee's 36 Video Game Learning Principles

The NRCR's five strands of mathematical proficiency were intended to establish guidelines for mathematics education—for what I am here calling *basic mathematics* education, though the NRCR did not use that term. The five strands were produced after considerable study and reflection, based on sound pedagogic principles. The primary target audience for the report was the United States school system, its teachers, and administrators. The words "video game" appear nowhere in the report.

Two years after *Adding It Up* was published, James Paul Gee, then a professor of reading at the University of Wisconsin, published his excellent and thought-provoking book *What Video Games Have to Teach Us About Learning and Literacy*. Professor Gee, who trained as a linguist, got into video games by trying to help his young son to play them. He discovered that not only were they quite hard, they seemed to be constructed around sound principles of learning. So he decided to make a study of them. In his book, Gee lists 36 principles of good education that can also be found in good video games. Although his book refers occasionally to math, mathematics education is not his focus. Rather, his 36 principles are the ones that video game designers unconsciously follow in order to produce compelling games that will sell to a fickle and demanding audience in a highly competitive market.

The NRC's *Adding It Up* and Gee's *What Video Games Have to Teach Us* are based on separate studies, grounded in different intellectual traditions, and aimed at different audiences. Yet, to anyone who reads these two books in close succession as I did, it is hard to believe that they were not the result of some collaboration, so extensive are the common themes and the agreement among them. If anything buttresses my point that video games are the ideal medium for learning basic mathematics, it is the degree to which the majority of Gee's 36 principles are exactly what it takes to achieve the five strands of the NRCR's mathematical proficiency. In this chapter, I'll present Gee's principles, and examine how they pertain to mathematics education. Before plunging into the details, however, I should answer the unspoken question that lies behind the incredulity that many people exhibit when they see the title of Gee's book.

Why Many Good Video Games Are All about Learning

In his book, Gee makes an important point that is often overlooked when the topic of video games comes up. I'll paraphrase his point here in my own terms. Video games are primarily a form of consumer entertainment. A video game succeeds in the marketplace only if enough people want to play it and enjoy playing it. A truly successful game will often entice people to play several hours a day for up to six months, sometimes longer. In the case of open-ended, exploration games like *World of Warcraft*, some people play for several years. (I'm not talking about fanatical players who arguably have an addiction to video games. They are the exceptions, although it is always the exceptions that make the news. I am a very busy university professor and have played *World of Warcraft* for many years now—though admittedly not daily and rarely for longer than a couple of hours at one sitting—and I still find it enjoyable.) It takes considerable effort on the part of the game designers to achieve this high level of commitment to voluntary involvement. Though there are now universities where you can learn the basic principles of game design and game engineering, for the most part game development has been to date very much intuition-driven, as each new game builds on what has been learned from previous ones. There have been some huge successes, but far more failures—some of them spectacular ones. As the industry has evolved, the leading designers have learned from experience what works and what does not. Even so, no one can predict in advance whether a particular game is going to be a success or not. As a result, game designers do massive amounts of user testing throughout the development process. This makes the process far more scientific and less speculative than the writing of a novel or the production of a movie.

In other words, game development has been a process of natural selection, or perhaps market selection would be a more accurate term. And in video game design as in nature, natural selection inevitably zeroes in on some significant factors. Since those factors concern human psychology, it should not be too surprising that, as Gee points out, many of the features that can be found in the successful video games are the principles of good education. The fact that many people do not recognize the educational features of video games reflects our society's established image of what education should be like, and to some extent the image of what video games are. The dominant image of education is, of course, the traditional school, with its desk-filled classrooms and the teacher at the front delivering pearls of wisdom, the receipt of which is supposed to forever change and benefit the recipient.

Okay, I admit I wrote that last sentence with a deliberate negative tone. There is a lot to be said for traditional classroom instruction. I spend a fair amount of my own time in such an environment, both as instructor and learner. But my point was to highlight that fact that the traditional classroom teaching model, while it has its good points, is by no means all there is to education. Many of the best students I encounter at the university figure out early in their educational careers that they can skip many—sometimes all—of the lectures and teach themselves using other sources. I did this myself when I was a student. (While possible in mathematics, I am not sure how effective it is in other subjects. And it's a dangerous strategy for all but the strongest and well-motivated student.)

As a species, humans are natural born learners. Every one of us. All it takes to stimulate someone to want to learn is interest or curiosity. Pick at random someone you think is "uneducated" and has "no interest in learning" and find out what they are particularly interested in. It could be pop music, baseball, naval battles, economics, politics, basket weaving, brewing beer, carpentry, sports, cooking, dogs—anything at all. Then ask them about that interest. You will find that they have a wealth of knowledge about it, and will constantly seek to learn more about it—though they may not call it learning.

The secret sauce that game developers stumbled across as they nosed their way forward to better and better games (in the sense of attracting greater numbers of players and retaining their interest) is to construct them as learning experiences, in the more general sense of the word "learning" I just described. This is not the only feature of a successful game, but it's a key one. Designers incorporate learning experiences not because they set out to construct "educational games." Their aim is to construct successful games that are fun to play. But because those games are sold to creatures that evolution has built to be inquisitive, curious, and instinctive learners—people—what emerges are products that have good learning principles built into them from the ground up.

Gee catalogued 36 of those learning principles, and in the remainder of this chapter I'll take a look at them and examine how each one pertains to mathematics education. I shall present them in a different order than Gee did, listing first the ones most relevant to mathematics, but I shall retain Gee's numbering.

Game Design Features Highly Relevant to Learning Basic Mathematics

> **Active, Critical Learning Principle**
> (Gee's Principle 1)
> *All aspects of the learning environment (including ways in which the semiotic domain is designed and presented) are set up to encourage active and critical, not passive, learning.*

I described the notion of a semiotic domain in Chapter 5. In the context of mathematics education, I think "reflective" would be a better adjective than Gee's "critical" here. As I've indicated previously, mathematics is not a spectator sport. It's about doing, not knowing. This is why an interactive environment such as a video game is in some ways much better for learning basic math than a textbook. I say "in some ways" because education is not a simple activity. At no point in this book do I claim that virtual worlds or video games are the only way people should learn math. What those technologies do is provide a powerful supplement to other forms of teaching that is ideally suited to learning basic math. A well-written textbook can be an excellent learning resource too—though as an educator I have observed that many textbooks are not well written and few students know how to make good use of them anyway.

> **Semiotic Domains Principle**
> (Gee's Principle 4)
> *Learning involves mastering, at some level, semiotic domains, and being able to participate, at some level, in the affinity group or groups connected to them.*

I discussed semiotic domains at some length in the previous chapter, and they are a major part of being able to do mathematics. This principle is closely related to the fifth of the NRCR's five strands of mathematical proficiency: productive disposition. School math instruction rarely seems to achieve it for many individuals, while video games often do. Whether that leads a player to start to think of him or herself as someone who is able to use mathematical ideas as they arise in the real world, even "at some level," remains to be seen.

"Psychological Moratorium" Principle
(Gee's Principle 6)
Learners can take risks in a space where real-world consequences are lowered.

A feature of mathematics that makes it psychologically difficult to learn is that in many cases, if a student fails to get the correct answer to a question, they are simply wrong. No half measures here; wrong is wrong. Given that degree of finality, it doesn't take many cases of being wrong to persuade some people to give up altogether, thinking they are simply not cut out for the subject. What they don't realize is that professional mathematicians are wrong almost all the time! If they are not, they are not working on hard enough problems, and are wasting their talents. This is not to say that professional mathematicians like failure. No one likes to fail. Rather, we mathematicians have learned that the path to eventual success is almost always littered with failures. The occasional successes are what make it worth the effort. The trick is to live with failure, to regard it as part of the process, and to learn from each mistake.

Of course, it's easy for someone like me to say this, right? I long ago got beyond the stage of having to take timed tests, where you don't get a second chance. I have the freedom to get something wrong and then try again. This points out one of the main problems with tests; they rarely allow for a second attempt, or a third, or however many it takes. They do not test if someone can do math; they determine if a person can do math within a specific period of time and get it right at the first (or if they work really fast, the second) attempt.

And it's not just timed tests that are the problem. Even the best intentioned classroom teacher cannot give all the students the time they need to complete a problem—to fail enough times to finally figure out how to do it. In most classrooms, there are students of very different abilities or different levels of achievement who work at different speeds, and the teacher has to settle for some middle ground that does not frustrate the faster students who find progress too slow or discourage the slower ones who cannot keep up. Needless to say, neither group comes away feeling well served; nor are they well served.

Classroom settings have another disadvantage. When a student gets a problem wrong, the teacher sees it and in many cases the rest of the class does too. For many people, failure is a bad thing—something to be avoided. In far too many classrooms (and other environments), public failure is far worse. It would be good for education to get beyond this state of affairs because when properly regarded, failure is an unavoidable part of learning and improvement. I don't see any chance of that happening on a widespread level. However, things will be very different for a player in a video game who is faced with solving a math problem, when Gee's sixth principle has a huge effect. Gee's next principle also becomes highly relevant in this regard.

Committed Learning Principle
(Gee's Principle 7)
Learners participate in an extended engagement (lots of effort and practice) as extensions of their real-world identities in relation to a virtual identity to which they feel some commitment and a virtual world that they find compelling.

When a video game player solves a problem, who should get the credit? After all, it is not exactly the player who is solving the problem, but the player's character in the game. Of course, the player controls the character, so .... There is a fascinating identity shift that goes on in video games that educators can leverage to great advantage, and a point I made earlier helps us understand how.

Apart from the dedicated first-person shooter games, most 3D video games give the player the option of playing first-person (they see the game world through their character's eyes) or with their character in full view. The most popular viewpoint, and the one that the game designers provide as the default, is for the player to be situated just behind and slightly above their character. In this position, the player easily identifies with the character (in *World of Warcraft*, I am very definitely my character). Yet because of this viewpoint, the player also has a sense of being outside the game (and thus outside the character) and controlling events—which is exactly what they are doing. This means that failure is not as personal as it is in real life; the player can mitigate failure by saying "It was my character who screwed up," even though the character failed only because the player did.

The distinction in player perception between first-person play and "view-from-above" play is very real, and manifests itself most dramatically in combat situations. Players who do battle in a game from the first-person perspective have a far greater sense of involvement and excitement than the view-from-above players, and this can be measured in terms of heart rate, breathing rate, and the various other physiological measures of excitement. Many players avoid the first-person perspective because it can become, well, just too scary, and defeat is too painful.

Of course, mathematics is not battling monsters, though for many students it comes pretty close on a metaphorical level. By trying to solve a math problem through their avatar, viewed from behind and above, players are a step removed from being the one who fails if they get it wrong. Of course, it is the player who solves, or fails to solve, the problem. (Although every player's character has built in artificial intelligence to move and to negotiate the world correctly and realistically, in no current games do avatars have any programmed mathematical problem solving abilities.) The players know this. Nevertheless, the sense of being a step removed from the game is very real, and that reduces the sense of failure when they get something wrong. Moreover, the players are anonymous, known only by their character's name unless they choose to let other players know their real identity. So if a player fails at some task, no one knows who it is.

Interestingly, despite the anonymity, the sense of social pressure on players of multiplayer games when they join guilds or team up in groups to complete tasks together is immense. Non-players such as parents or spouses generally find it beyond belief when a player says, "Sorry, I can't come to dinner just now; we're deep into a dungeon and I can't leave my group, they depend on me." But for online gamers, that sense of obligation is every bit as real and compelling as in real life. One thing that does transfer almost unchanged from the real world to multiplayer games is social behavior. This is true even when the player has never met the others in person, and perhaps knows nothing about their real-life identities. For the designer of a multiplayer game intended to meet a mathematical education goal, the situation could not be better. The lack of complete identification with the character, together with the anonymity, means that the sense of failure is reduced, though not eliminated, but that's a huge advantage. And the strong social pressures mean that players can work together and for each other in a group, and will stick to it until the problem is solved.

Of course, mathematical tasks are not the only ones that game players fail at. Progression through a video game is all about exploration, about being continually tested, about trial and error, and about success and failure. There is no manual. (The booklet that usually comes with the game and is titled a "manual" is really just a description of how to operate the game mechanics. Gamers hardly ever read manuals; they just play the game and figure things out as they go. Game designers know this and design the game accordingly.) When video games first started to appear, game designers faced the same problem that math teachers do: how do you ensure that players do not give up after they have failed several times to complete a particular task? It took a long time to get it right.

In many video games, when you really screw up, your character gets killed. But obviously, the game designer did not want you to stop playing at that point. Thus, "death" in a video game is not a permanent thing; there is always the option of resurrection. In early games, this often meant you started out afresh, or at least you started out at the beginning of some previously attained threshold level. In addition to having to go through the whole process again—and that could require several hours of effort—in those early games you also lost all of the money and items you had just acquired. This meant that although death was not permanent, the cost of being killed was high. The result was that many players adopted an extremely cautious approach, which made the game seem boring and so they gave up. The players that played more aggressively kept getting killed and having to repeat several hours of play, and they too eventually got frustrated and gave up. In both cases the outcome was a lost player. And, particularly in an era of online chat and blogging, lost players rapidly turn into losses of future sales, as other gamers learn of the frustrations of their friends. So the game designers kept adjusting the formula—the cost of dying—until they found one that worked well. Extremely well, in fact.

Here is how many current video games (such as *World of Warcraft*) handle character death. When a character dies, the character's ghost is transported to a nearby "graveyard" where the player has the option of paying game money, or some other game asset, to be brought back to life immediately by some god-like entity, or the player can run back to their corpse where the character died and resurrect themselves at little or no cost. There is a real cost to the player who elects self-resurrection: time. It can take anything from a few seconds to several minutes to run back to the location of the corpse, and during that time the player generally cannot interact with the game world in any way except to direct the motion of the character. By tweaking the various costs associated with resurrecting a dead character—game money, game assets, or time—game designers came up with a formula that encouraged players to be bold and to try out new ideas, yet provided a strong incentive to get it right, if not at the first try, then as soon as possible thereafter, and not get killed too often. In other words, it is to the player's advantage both to act smart and to take calculated risks.

I should also note that one major advantage the designer of an online game has over the designer of an out-of-the-box, offline video game is that the online game provider can continuously monitor game play (and all of the blog sites and bulletin boards devoted to the game that inevitably spring up) and keep adjusting—by way of upgrade patches downloaded over the Internet the next time the player logs on—the appropriate parameters in order to optimize the game experience.

Self-Knowledge Principle

(Gee's Principle 9)

The virtual world is constructed in such a way that learners learn not only about the domain but about themselves and their current and potential capacities.

This is essentially one of the NRCR's five strands of mathematical proficiency we discussed in the previous chapter: productive disposition. The NRCR described this as: "habitual inclination to see mathematics as sensible, useful, and worthwhile, combined with a confidence in one's own ability to master the material."

Amplification of Input Principle
(Gee's Principle 10)
For a little input, learners get a lot of output.

Good teachers have always known that students perform much better when their efforts lead to positive results. No one likes to see their work disappear into a black hole with no sign of any effect. The greater the positive output for a given input, the greater the incentive to repeat that kind of input.

In playing *World of Warcraft*, I almost gave up half way through the 40s levels. By the time a player reaches level 30, some of the challenges are quite difficult, but they lead to advancement in the game such as leveling up and desirable drops from kills. But somewhere in the mid-40s, progress seems to slow. The ratio of output to input drops, or so it seems. I know I am not alone in having had that feeling. This common player reaction may have been accidental on the part of the *WoW* level designers. On the other hand, I did not stop playing, and looking back on my entire progression from level 1 to level 60 (the top level prior to the release of the first expansion pack in January 2007), the slow progress through the 40s did add variety to the overall experience, so maybe the level designers knew what they were doing by setting the stage for an exciting progression from level 50 to level 60. After all, no matter how exciting our careers or leisure pursuits may be, real life also has its slow, boring periods. Without lows there can be no highs.

Regardless of the game structure of *WoW*, for any video game designed to teach math, my suspicion is that it would be a mistake to allow a stretch of the game when there is no alternative to slow progress. There is already a danger of players giving up because they find the math challenges too hard (though the game can be engineered to enable them to continue playing without solving the math challenges), and doing the math parts is already likely to slow down progress. To counter this, a video game designer should take note of Gee's Amplification of Input principle. Coupled with his Achievement Principle below, this can ensure that completion of a math challenge brings a significantly better reward than doing a non-math challenge.

Achievement Principle
(Gee's Principle 11)
For learners of all levels of skill there are intrinsic rewards from the beginning, customized to each learner's level, effort, and growing mastery, and signaling the learner's ongoing achievements.

All good teachers know this, but in a normal classroom setting it is almost impossible to achieve. Video games provide a perfect environment to meet this learning principle because each player's game experience is unique, and he or she is the center of (the "star" in) their own progress through the game. In particular, although in principle mathematics is about solving the problem, and not about solving it within a given time frame, our current education system all too often turns math into a race. Students who are naturally slower than others—and many world-class mathematicians fall into this category—frequently become discouraged and decide that they are "not good at math." In a classroom setting, it's actually hard not to impose some time limits on the students, but in a video game each student can progress at his or her own pace.

This is not to say that time is completely irrelevant in teaching math. Important parts of learning math include having automatic recall of basic facts such as the multiplication tables, and what is known as "chunking," achieving automatic fluency in particular sequences of steps. The only way to achieve either of these is by repetitive practice under tighter and tighter time limits. Video games offer a huge advantage in this regard, in that they can make repetitive actions under time pressure fun. It may sound weird, I know, but they do, and this principle works together with Gee's 12th principle.

Practice Principle
(Gee's Principle 12)
Learners get lots and lots of practice in a context where the practice is not boring (i.e., in a virtual world that is compelling to learners on their own terms and there the learners experience ongoing success). They spend lots of time on task.

In mathematics as in most other walks of life, practice makes perfect. In fact, practice not only makes perfect, it's what it takes to become competent. One of the principal reasons why many students give up on math is that they don't practice enough. It's not that they are not able to practice anything. The kid who blows off his or her math homework might spend several hours struggling to master a particular skateboard maneuver or guitar riff. What keeps them plugging away at those activities in the face of one failed attempt after another? It doesn't take rocket science to come up with the answer. They keep trying because they can see the final outcome, and they want to achieve it. The trouble with math, at least the way it is typically presented in today's education system, is that the students have no idea what the outcome might be. In essence, the most they hear by way of motivation is all too often, "This is important stuff that will be useful to you some day, so you should work at it." Would that convince you? Sometimes it might, especially if you really admired and trusted the person saying it and recognized that they knew what they were talking about. But in general, I am sure you'd want more by way of motivation before you'd be willing to give up a pile of your time—to give up doing other activities that interest you—for something that has no intrinsic appeal to you.

A fascinating feature of video games is how they make repetitive actions seem fun. If a player has to spend hours repeating a particular task to advance in the game, why not endow that task with a particular educational purpose? Learning fast, automatic recall of the multiplication table, for example. This is feasible for a mathematical activity that has been designed to arise naturally in the game. For a mathematical activity that does not fit naturally in the game, using game dynamics to get players to practice the activity under time pressure may or may not work. There are some video games on the market that adopt just this approach (e.g., *Timez Attack*, by the company Big Brainz), and the anecdotal evidence is that they work remarkably well.

"Regime of Competence" Principle
(Gee's Principle 14)
The learner gets ample opportunity to operate within, but at the outer edge of, his or her resources, so that at those points things are felt as challenging but not "undoable."

This principle is a crucial factor in designing a successful video game. It's why the level designer is such a key person in game development. It's also a key feature of good teaching. Students must advance and yet not be presented with a challenge that is so far beyond their current ability that they become discouraged and give up. Good teachers know this and put together lesson plans accordingly.

But there is a crucial difference between a math teacher and a video game level designer. A teacher who does a poor job of classroom "level design" generally does not lose his or her job. But a video game that players find either too easy or else impossibly hard will not sell; players will not play it—and the level designer might well be out of a job, very likely along with everyone else at the company. The commercial video game market is a much harder taskmaster than the school system.

My experience playing *World of Warcraft* is that frequently I feel that I cannot complete a particular task; either my armor (my main character is a warrior) is not good enough, or my weapons are inadequate, or I cannot figure out what sequence of steps to take. But on every such occasion, it is not too long before something happens in the game to help me succeed. The game designers made sure that I am not left dangling for too long. Admittedly, this means that my sense of "facing a difficult challenge" is a bit of a fake. For all that I feel I am battling forward valiantly on my own, with only my wits to guide me, in reality I am following a fairly constrained path that the game designer has set up for me. But that is exactly what a lot of good education is about. The student should feel that he or she has solved the problem or made the discovery on his or her own, but in reality the teacher should present the student only with tasks and problems that with first or subsequent efforts are doable for that student, and provide the resources the student requires. The feeling that a player of a well-designed game gets is one of constantly pushing just beyond what seems possible. The designer of a mathematics educational game should set out to achieve the same dynamic, the difference being that some of the challenges will involve solving (naturally arising) math problems.

Probing Principle
(Gee's Principle 15)
Learning is a cycle of probing the world (doing something); reflecting in and on this action and, on this basis, forming a hypothesis; re-probing the world to test this hypothesis; and then accepting or rethinking the hypothesis.

Again, all good video games exhibit this feature. So does good education, though too often it is missing in the classroom. Students who simply sit in a classroom and listen to the teacher "instructing" may acquire some facts—though there are far more efficient ways to soak up knowledge—but true learning involves action on the part of the student. This is particularly true for mathematics, which, as I keep emphasizing, is about doing, not about knowing. Indeed, Gee's 15th principle is a perfect description of learning to do math!

Multiple Routes Principle

(Gee's Principle 16)

There are multiple ways to make progress or move ahead. This allows learners to make choices, rely on their own strengths and styles of learning and problem solving, while also exploring alternative styles.

The days are long gone when math teachers used to insist that everyone had to solve a problem the same way (often claiming that in math there is always only one correct way, a ludicrous statement that spoke volumes about the knowledge and psychological make-up of the teacher). But even the most gifted and informed teacher, when faced with a state-mandated exam at the end of the year, has difficulty allowing too much freedom for exploration in a class of 20 or 25 students. In contrast, video games are designed around this form of learning. A good math ed game should provide it in spades.

In designing such a game, the designer could build in the possibility that when a player is unable to solve a particular math problem, he or she will not be held up in the game, but can put that problem aside and go on to face different kinds of challenges. This will allow the player to continue to advance in the game. To encourage the player to keep coming back and trying the math problem in different ways, the designer could make sure that the only way to get certain particularly desirable game rewards is by solving the math problems. The idea is that, in the game a player can obtain the coolest rewards only by solving math problems. *World of Warcraft* provides just such a dual-path structure.

Additionally, the designers of *World of Warcraft* clearly wanted players to cooperate in groups. Many of the most attractive items can be obtained only by completing dungeon quests—tasks that generally require several players to act in a coordinated fashion. It can take three or more hours of continuous play to complete such a mission. Some players, myself included, rarely have so much continuous time at our disposal, and so we solo our way through the game, perhaps cooperating with another player for a few minutes now and then on an ad hoc basis, but for the most part playing alone. I was able to solo my way all the way up to level 70 and beyond, but for the most part I had to forgo doing those long dungeon quests and getting my hands on that really cool equipment.

Situated Meaning Principle
(Gee's Principle 17)
The meanings of signs (words, actions, objects, artifacts, symbols, texts, etc.) are situated in embodied experience. Meanings are not general or decontextualized. Whatever generality meanings come to have is discovered bottom up via embodied experiences.

This is why the learning that takes place in video games does not seem like learning. It is what the researchers saw in the street market in Recife.

Text Principle
(Gee's Principle 18)
Texts are not understood purely verbally (i.e., only in terms of the definitions of the words in the text and their text-internal relationships to each other) but are understood in terms of embodied experiences. Learners move back and forth between texts and embodied experiences. More purely verbal understanding (reading texts apart from embodied action) comes only when learners have had enough embodied experience in the domain and ample experiences with similar texts.

This principle underlies the insistence by many in the field of mathematics education that it is essential that students gain conceptual understanding—that they know what all those symbols and procedures *mean*. But notice that Gee does not claim that purely symbolic, rule-based learning is not possible. Rather, his point is that either component on its own doesn't get you very far.

For most people, a key ability they need in order to live a fruitful and successful life is a reasonable competency in basic mathematical reasoning as it arises in real-world situations. If they can also come to grips with abstract, symbolic math, then so much the better. But in terms of everyday life, everyone needs (embodied) basic math skills, while only those who choose to follow certain career paths actually need symbolic math skills. As those who do go on to master some abstract mathematics invariably discover, making progress requires a constant back and forth between understanding and learning to follow rules that do not yet make sense.

Intertextual Principle
(Gee's Principle 19)

The learner understands texts as a family ("genre") of related texts and understands any one such text in relation to others in the family, but only after having achieved embodied understandings of some texts. Understanding a group of texts in a family (genre) of texts is a large part of what helps the learner make sense of such texts.

This is to some degree an extension of the previous principle. In the case of mathematics, which is arguably a special case, the "genres" would be the different branches of math.

Multimodal Principle
(Gee's Principle 20)

Meaning and knowledge are built up through various modalities (images, texts, symbols, interactions, abstract design, sound, etc.), not just words.

Remember, the written symbols that people often identify with mathematics are merely *representations* of mathematics. They are no more mathematics than musical notation is music. Gee's 20th principle applies completely to mathematics, particularly if we read his final clause "not just words" as "not just words and math symbols." One of the advantages of virtual worlds is that there are many ways to facilitate the student constructing mathematical meaning and knowledge—which ultimately are meaning and knowledge of the real world.

Incremental Principle
(Gee's Principle 24)

Learning situations are ordered in the early stages so that earlier cases lead to generalizations that are fruitful for later cases. When learners face more complex cases later, the learning space (the number and type of guesses the learner can make) is constrained by the sorts of fruitful patterns or generalizations the learner has found earlier.

This is an obvious feature of both good video games and good teaching. It is one of the reasons why learning in a suitably constructed video game in a virtual world can actually be more effective than learning in the real world, say at a street market stall, and why the game structure (which orders the player's experiences) can be of great importance.

But note that Gee's principle applies only to the early stages of the game. This restriction is particularly important in a game to develop mathematical thinking because the learner needs to learn how to cope with a novel problem for which she or he has not seen a simpler case. The more advanced student, at a higher level in the game, needs to develop a strategy of generating her or his own simpler cases in order to develop techniques to tackle the original problem.

Concentrated Sample Principle
(Gee's Principle 25)
The learner sees, especially early on, many more instances of fundamental signs and actions than would be the case in a less controlled sample. Fundamental signs and actions are concentrated in the early stages so that learners get to practice them often and learn them well.

This is another reason why video games in virtual worlds can have a learning advantage over real-world situations for teaching math. In the real world, you have to take things as they come at you.

Bottom-Up Basic Skills Principle
(Gee's Principle 26)
Basic skills are not learned in isolation or out of context; rather, what counts as a basic skill is discovered bottom up by engaging in more and more of the game/domain or game/domains like it. Basic skills are genre elements of a given type of game/domain.

This principle is important for the same reasons as the previous one.

Explicit Information On-Demand and Just-in-Time Principle
(Gee's Principle 27)
The learner is given explicit information both on-demand and just-in-time, when the learner needs it or just at the point where the information can best be understood and used in practice.

This is a huge educational factor that we shall return to later. Video games have it down to a fine art. Recall my earlier remarks about my subjective feelings playing *World of Warcraft*. This provides yet another advantage video games have over the real world when it comes to education.

Discovery Principle
(Gee's Principle 28)
Overt telling is kept to a well-thought-out minimum, allowing ample opportunity for the learner to experiment and make discoveries.

As the old saying goes, "You tell me, I forget; you show me, I remember; you let me discover, and I know." Good teachers try to let their students encounter things for themselves, but in a classroom setting, with the pressures of statewide exams, there is a strict limit on the amount of exploration that can go on. In the classroom, there is also the possibility of one student doing most of the work and the others simply copying, or at least taking the lead from that student. Video games allow each student to explore at his or her own pace.

Transfer Principle
(Gee's Principle 29)
Learners are given ample opportunity to practice, and support for, transferring what they have learned earlier to later problems, including problems that require adapting and transforming that earlier learning.

Note that this is not the same notion of "transfer" that educators typically use, which is to refer to taking something learned in one context and applying it in another. Rather, Gee's meaning is closer to what mathematics educators call abstraction (or in one case "vertical mathematization"). According to Jean Lave, true context-to-context transfer is not really possible because everything is situated. In mathematics effective transfer is made possible by the process of abstraction, which allows us to look for relations and structures that are similar.

However, some research shows that it takes many experiences before the situated insights are recognized as general tools that can be used in future situations. In other words, we have to become skilled at looking for familiar structures as structures. Context-to-context transfer is impossible without this shift to recognizing structural similarity, and this is an important part of learning to think mathematically.

Many teachers try to achieve Gee's transfer principle in their classrooms, but given the need for lots of experience to maximize its application, the reality is there is insufficient time available for students to acquire this ability. Video games, with their player-centric structure that allows for repeated exposure and many repetitions with full student engagement, are much better equipped to provide this form of learning.

Dispersed Principle
(Gee's Principle 34)
Meaning/knowledge is dispersed in the sense that the learner shares it with others outside the domain/game, some of whom the learner may rarely or never see face-to-face.

One of the positive outcomes a multiplayer game can provide is that players come to view mathematical ability as something they share with others—a body of knowledge and skills that cuts across all nations and people on Earth, of any race, gender, ethnicity, or culture. Another valuable outcome is a much wider acceptance of and appreciation for mathematics, and maybe a video game can help bring this about too.

Affinity Group Principle
(Gee's Principle 35)
Learners constitute an "affinity group," that is, a group that is bonded primarily through shared endeavors, goals, and practices and not shared race, gender, nation, ethnicity, or culture.

This principle is important for the same reasons as the previous one.

Insider Principle

(Gee's Principle 36)

The learner is an "insider," "teacher," and "producer" (not just a "consumer") able to customize the learning experience and domain/game from the beginning and throughout the experience.

This is a vitally important feature of good education in any discipline. Since basic math is about how each individual person encounters and deals with the world, the most appropriate and productive way for people to learn it is by taking control of their own learning in a suitably structured framework. Video games do this by their very nature.

Game Design Features Moderately Relevant to Learning Basic Mathematics

I turn now to the Gee principles that are—as far as we know today—only moderately relevant to learning basic mathematics.

Semiotic Principle

(Gee's Principle 3)

Learning about and coming to appreciate interrelations within and across multiple sign systems (images, words, actions, symbols, artifacts, etc.) as a complex system is core to the learning experience.

In the case of mathematics, it would be easy (and correct) to read this as referring to abstract, symbolic math, but that would be to miss the point as far as learning math in a video game is concerned. As we have already observed, while highly suited to learning basic math, video games seem poorly suited for learning advanced, symbolic math. It's likely that a video game could motivate students—players—to practice symbolic math, but until someone tries it we won't know whether or how well it could work. But notice what Gee has to say about "sign systems" such as "images, words, actions, symbols, artifacts, etc." Part of learning mathematics is recognizing that it all fits together—that it is not just a collection of unconnected techniques. One way to address this in a video game is to repeatedly present players with different representations of the same concept; for example, the relationship between fractions, proportions, decimals, and percentages.

Metalevel Thinking about Semiotic Domains Principle
(Gee's Principle 5)

Learning involves active and critical thinking about the relationships of the semiotic domain being learned to other semiotic domains.

This principle raises the same issues as the previous one.

Ongoing Learning Principle
(Gee's Principle 13)

The distinction between learner and master is vague, since learners, thanks to the operation of the "regime of competence" principle listed [as principle 14], must, at higher and higher levels, undo their routinized mastery to adapt to new or changed conditions. There are cycles of new learning, automatization, undoing automatization, and new reorganized automatization.

In terms of mathematics education in general, this principle is extremely relevant. The reason I have classified it as "moderately relevant" is that the real power of video games is in helping students learn everyday math. And in everyday math, there are simply not enough different stages or levels of learning for Gee's 13th principle to apply significantly in a game setting. If video games were developed to cover more advanced parts of mathematics, the designers would certainly have to take this principle into account.

Subset Principle
(Gee's Principle 23)

Learning even at its start takes place in a (simplified) subset of the real domain.

This principle certainly applies to learning math. I relegated it to the "moderately relevant to learning mathematics" category because it is not a principle to pay conscious attention to in designing a video game to teach math. The reason is that mathematics is itself already a simplified subset of the real domain. Indeed, I would say that it is the ultimate simplified subset, as incredible as this may seem to those who find mathematics anything but simple.

Game Design Features Not Particularly Relevant to Learning Mathematics

Finally, I list these mainly to provide compete coverage of Gee's principles. Having said that, I note that any video game satisfies all of Gee's principles. Mathematics is not something that exists in isolation, and no one would seriously suggest that education in any one subject should be at the expense of education in another. Thus, other forms of learning should, and will, take place in any game, with some of that additional learning by virtue of the Gee principles listed below.

Design Principle
(Gee's Principle 2)
Learning about and coming to appreciate design and design principles is core in the learning experience.

Identity Principle
(Gee's Principle 8)
Learning involves taking on and playing with identities in such a way that the learner has real choices (in developing the virtual identity) and ample opportunity to meditate on the relationship between new identities and old ones. There is a tripartite play of identities as learners relate, and reflect on, their multiple real-world identities, a virtual identity, and a projective identity.

"Material Intelligence" Principle
(Gee's Principle 21)
Thinking, problem solving, and knowledge are "stored" in material objects and the environment. This frees learners to engage their minds with other things whole combining the results of their own thinking with the knowledge stored in material objects and the environment to achieve yet more powerful effects.

Intuitive Knowledge Principle
(Gee's Principle 22)

Intuitive or tacit knowledge built up in repeated practice and experience, often in association with an affinity group, counts a great deal and is honored. Not just verbal and conscious knowledge is rewarded.

Cultural Models about the World Principle
(Gee's Principle 30)

Learning is set up in such a way that learners come to think consciously and reflectively about some of their cultural models regarding the world, without denigration of their identities, abilities, or social affiliations, and juxtapose them to new models that may conflict with or otherwise relate to them in various ways.

Cultural Models about Learning Principle
(Gee's Principle 31)

Learning is set up in such a way that learners come to think consciously and reflectively about their cultural models of learning and themselves as learners, without denigration of their identities, abilities, or social affiliations, and juxtapose them to new models of learning and themselves as learners.

Cultural Models about Semiotic Domains Principle
(Gee's Principle 32)

Learning is set up in such a way that learners come to think consciously and reflectively about their cultural models about a particular semiotic domain, without denigration of their identities, abilities, or social affiliations, and juxtapose them to new models about this domain.

Distributed Principle
(Gee's Principle 33)

Meaning/knowledge is distributed across the learner, objects, tools, symbols, technologies, and the environment.

Developing Mathematical Proficiency in a Video Game

Having listed the five strands of mathematical proficiency in Chapter 6, I'll now look in turn at how teachers can take advantage of a suitable video game (when they become available) to help students acquire mastery of each strand, and what game developers can do to facilitate that learning. I start with conceptual understanding.

How Important Is Conceptual Understanding?

First, I'll note that full conceptual understanding, while desirable, is not strictly necessary in order to be able to apply mathematics successfully. Differential calculus was first developed in the middle of the seventeenth century, and has been used extensively and successfully ever since. Yet it was not until the very end of the nineteenth century, some 250 years later, that mathematicians achieved full conceptual understanding, and then only after considerable effort had been exerted to develop the mathematical machinery necessary to ground such understanding. Even today, the experience most students have when they learn calculus is that it's possible to ace all the calculus exams without understanding what calculus is or how it works. That was definitely how I learned calculus. And I know from having included on my calculus exams for many years at least one question that tries to elicit from my students their degree of understanding, that almost all of them don't understand it any more than I did at that stage. So much for the much-touted myth that doing mathematics requires conceptual understanding.

What mathematical thinking does require, I suggest, is *functional understanding*. Let me explain what I mean by that term. I'll start out with a question:

what exactly does it mean to understand an abstract mathematical concept? Take the most familiar example of all: positive whole numbers. What does it mean to say that a child (or adult for that matter) has mastered the concept of a positive whole number? This question is not as clear-cut as might first be assumed. Moreover, we can appreciate just what a major cognitive leap it is to grasp this purely abstract concept by recalling that it took many generations to develop; early humans were able to count collections long before numbers came on the scene.

The Number Concept

Numbers,[1] specifically whole numbers, arise from the recognition of patterns in the world around us; the pattern of *oneness*, the pattern of *twoness*, the pattern of *threeness*, and so on. To recognize the pattern that we call *threeness* is to recognize what it is that a collection of three apples, three children, three footballs, and three rocks have in common. "Can you see a pattern?" a parent might ask a small child, showing her various collections of objects—three apples, three shoes, three gloves, and three toy trucks. The counting numbers 1, 2, 3, and so on are a way of capturing and describing those patterns. The patterns captured by numbers are abstract, and so are the numbers used to describe them.

At the age of about five or younger, the typical child in an educated, Western culture makes a cognitive leap that took humankind many thousands of years to achieve: the child acquires the concept of number. She comes to realize that there is something common to a collection of, say, five apples, five oranges, five children, five cookies, a rock group of five members, and so on. That common something, *fiveness*, is somehow captured or encapsulated by the number 5, an abstract entity that the child will never see, hear, feel, smell, or taste, but which will have a definite existence for the rest of her life. Indeed, such is the role numbers play in everyday life that, for most people, the ordinary counting numbers 1, 2, 3 . . . are more real, more concrete, and certainly more familiar, than Mount Everest or the Taj Mahal.

It should be noted that a typical child learns to count, i.e., to recite the number words "one, two, three, four, five" before she learns to use them to count collections of objects. In keeping with one of my general points that as symbolic creatures we can learn to play language games, children learn the numbers first and then realize that they can use them to determine the size of a collection. People who claim that understanding must precede learning the rules in mathematics education seem to overlook the well-known fact that it doesn't work that way even at the very first stage of learning to count. For young children, counting begins as

---

[1] This section is adapted from Keith Devlin, *The Language of Mathematics: Making the Invisible Visible*, W. H. Freeman, 1998.

a meaningless game—though one that they seem to enjoy—and only later does the child come to understand the number concept and be able to use numbers to count collections.[2]

The conceptual creation of the counting numbers marks the final step in the process of recognizing the pattern of "number of members of a given collection." This pattern is completely abstract; indeed, so abstract that it is virtually impossible to talk about it except in terms of the abstract numbers themselves. Try explaining what is meant by a collection of 25 objects without referring to the *number* 25. (With a very small collection, you can make use of your fingers; a collection of five objects can be explained by holding up the fingers of one hand and saying, "This many.")

The acceptance of abstraction does not come easily to the human mind. Given the choice, people prefer the concrete to the abstract. Indeed, work in psychology and anthropology indicates that a facility with abstraction seems to be something we are not born with but acquire, often with great difficulty, as part of our intellectual development. For instance, according to the work of the cognitive psychologist Jean Piaget, the abstract concept of volume is not innate, but rather is learned at an early age. Young children are not able to recognize that a tall, thin glass and a short, stout one can contain the same volume of liquid, even if they see one poured into the other. For a considerable time, they will maintain that the quantity of liquid changes and that the tall glass contains more than the short one.

Piaget also found that young children who can "count" think that there are more objects when they are spread out than there are if the same number of objects are grouped closely together. This suggests they do not yet have conceptual understanding of a natural number. The concept of abstract number seemingly has to be learned, and small children appear to acquire this concept *after* they have learned to "count." Evidence that the concept of number is not innate comes from the study of cultures that have evolved in isolation from modern society. For instance, when a member of the Vedda tribe of Sri Lanka wants to count coconuts, he collects a heap of sticks and assigns one to each coconut. Each time he adds a new stick, he says, "That is one." But if asked to say how many coconuts he possesses, he simply points to the pile of sticks and says, "That many." The tribesman thus has a type of counting system, but far from using abstract numbers, he "counts" in terms of decidedly concrete sticks. The Vedda tribesman employs a system of counting that dates back to very early times, that of using one collection of objects, say sticks or pebbles, to count the members of another collection, by pairing off the sticks or pebbles with the objects to be counted.

---

[2] See, for example, Terezinha Nunes and Peter Bryant, *Children Doing Mathematics (Understanding Children's Worlds)*, Blackwell, 1996.

The Origin of the Number Concept

The earliest known man-made artifacts believed to be connected to counting are notched bones, some of which date back to around 35,000 BCE. At least in some cases, the bones seem to have been used as lunar calendars, with each notch representing one sighting of the moon. Similar instances of counting by means of a one-to-one correspondence appear again and again in preliterate societies; pebbles and shells were used in the census in early African kingdoms, and cacao beans and kernels of maize, wheat, and rice were used as counters in the New World. Of course, any such system suffers from an obvious lack of specificity. A collection of notches, pebbles, or shells indicates a quantity but not the kinds of items being quantified, and hence cannot serve as a means of storing information for long periods.

The first known enumeration system that solved this problem was devised in what is now the Middle East, in the region known as the Fertile Crescent, stretching from present-day Syria to Iran. During the 1970s and early 1980s, anthropologist Denise Schmandt-Besserat of the University of Texas at Austin carried out a detailed study of clay artifacts found in archaeological digs at various locations in the Middle East. At every site, among the usual assortment of clay pots, bricks, and figurines, Schmandt-Besserat noticed the presence of collections of small, carefully crafted clay shapes, each measuring between one and three centimeters across: spheres, disks, cones, tetrahedrons, ovoids, cylinders, triangles, rectangles, and the like. The earliest such objects dated back to around 8,000 BCE, some time after people started to develop agriculture and they first needed to plan harvests and lay down stores of grain for later use.

An organized agriculture required a means of keeping track of a person's stock, and a means to plan and to barter. The clay shapes examined by Schmandt-Besserat appear to have been developed to fulfill this role, with the various shapes being used as tokens to represent the kind of object being counted. For example, there is evidence that a cylinder stood for an animal, cones and spheres stood for two common measures of grain (approximately a peck and a bushel, respectively), and a circular disk stood for a flock. In addition to providing a convenient, physical record of a person's holdings, the clay shapes could be used in planning and bartering, by means of physical manipulation of the tokens.

By 6,000 BCE, the use of clay tokens had spread throughout the region. Over time, with the increasingly complex societal structure of the Sumerians—characterized by the growth of cities, the rise of the Sumerian temple institution, and the development of organized government—more elaborate forms of token were introduced. These newer tokens had a greater variety of shapes, including rhomboids, bent coils, and parabolas, and were imprinted with markings. Whereas the plain tokens continued to be used for agricultural accounting, these more complex tokens appear to have been introduced to represent manufactured objects such as garments, metalworks, jars of oil, and loaves of bread.

The stage was set for the next major step toward the development of abstract numbers. Around 3000 BCE, as state bureaucracy grew, two means of storing the clay tokens became common. The more elaborate, marked tokens were perforated and strung together on a string attached to an oblong clay frame, and the frame was marked to indicate the identity of the account in question. The plain tokens were stored in clay containers, which were hollow balls some five to seven centimeters in diameter, and the containers were marked to show the account. Both the strings of tokens and the sealed clay envelopes of tokens thus served as accounts or contracts. Of course, one obvious drawback of a sealed clay envelope is that the seal has to be broken open in order to examine the contents. So the Sumerian accountants developed the practice of impressing the tokens on the soft exteriors of the envelopes before enclosing them, thereby leaving a visible exterior record of the contents. But with the contents of the envelope recorded on the exterior, the contents themselves became largely superfluous—all of the requisite information was stored on the envelope's outer markings. The tokens themselves could be discarded, which is precisely what happened after a few generations. The result was the birth of the clay tablet, on which impressed marks, and those marks alone, served to record the data previously represented by the tokens. In present-day terminology, we would say that the Sumerian accountants had replaced the physical counting devices by written *numerals.*

From a cognitive viewpoint, it is both interesting and highly relevant to mathematics education that the Sumerians did not immediately advance from using physical tokens sealed in a marked envelope to using markings on a single tablet, because it indicates the difficulty of the mental leap involved. For some time, the marked clay envelopes redundantly contained the actual tokens depicted by the outer markings. The tokens were regarded as representing the quantity of grain or the number of sheep, while the envelope's outer markings were regarded as representing not the real-world quantity but the tokens in the envelope. That it took so long to recognize the redundancy of the intermediate tokens suggests that going from physical tokens to an abstract representation was a considerable cognitive development.

Of course, the adoption of a symbolic representation of the amount of grain does not in itself amount to the explicit recognition of the number concept in the sense familiar today, where numbers are regarded as *things*, as *abstract objects*. Exactly when humankind achieved that recognition is hard to say, just as it is not easy to pinpoint the moment when a small child makes a similar cognitive advance. What is certain is that, once the clay tokens had been abandoned, the functioning of Sumerian society relied on the notions of oneness, twoness, threeness, and so on, since that is what the markings on their tablets denoted.

Arithmetic

Having a number concept and using that concept to count, as the Sumerians did, is one thing; having a conceptual understanding of numerical procedures such as addition (the most basic arithmetical operation) is quite another. This latter development came much later, when people first began to carry out intellectual investigations of the kind that we would now classify as mathematics.

As an illustration of the distinction between the use of a mathematical device and the explicit recognition of the entities involved in that device, take the familiar observation that order is not important when a pair of counting numbers (natural numbers) is added or multiplied. Using modern algebraic terminology, this principle can be expressed in a simple, readable fashion by the two commutative laws:

$$m + n = n + m \quad \text{and} \quad m \times n = n \times m.$$

In each of these two identities, the symbols m and n are intended to denote any two natural numbers. Using these symbols is quite different from writing down a particular instance of these laws: for example,

$$3 + 8 = 8 + 3 \quad \text{and} \quad 3 \times 8 = 8 \times 3.$$

The second case is an observation about the addition and multiplication of two particular numbers. It requires our having the ability to handle individual abstract numbers, at the very least the abstract numbers 3 and 8, and is typical of the kind of observation that was made by the early Egyptians and Babylonians. But it does not require a well-developed concept of abstract number or the abstract concept of addition, as do the commutative laws.

By around 2000 BCE, both the Egyptians and the Babylonians had developed primitive numeral systems. They knew addition and multiplication were commutative, in the sense that they were familiar with these two patterns of behavior, and undoubtedly made frequent use of commutativity in their daily calculations. But in their writings, when describing how to perform a particular kind of computation, they did not use algebraic symbols such as m and n. Instead, they always referred to particular numbers, although it is clear that in many cases the particular numbers chosen were presented purely as examples, and could be freely replaced by any other numbers.

For example, in the so-called Moscow Papyrus, an Egyptian document written in 1850 BCE, there appear the following instructions for computing the volume of a certain truncated square pyramid (one with its top chopped off by a plane parallel to the base):

> If you are told: a truncated pyramid of 6 for the vertical height by 4 on the base by 2 on the top. You are to square this 4, result 16. You are to double 4, result 8. You are to square 2, result 4. You are to add the 16, the 8, and the 4, result 28. You are to take a third of 6, result 2. You are to take 28 twice, result 56. See, it is 56. You will find it right.

Though these instructions are given in terms of particular dimensions, they clearly only make sense as a set of instructions if the reader is free to replace these numbers by any other appropriate values. In modern notation, the result would be expressed by means of an algebraic formula; if the truncated pyramid has a base of sides equal to a, a top of sides equal to b, and a height h, then its volume is given by the formula

$$V = (1/3)h(a^2 + ab + b^2).$$

Being aware of and utilizing a certain pattern is not the same as formalizing that pattern. The commutative laws, for example, express certain patterns in the way the natural numbers behave under addition and multiplication, and the laws express these patterns in an explicit fashion. By formulating the laws using algebraic indeterminates such as m and n, entities that denote arbitrary natural numbers, we place the focus on the *concept* of addition or multiplication, not some particular instance of an addition or *a* multiplication. The general concepts of addition and multiplication were not developed until the era of ancient Greek mathematics began around 600 BCE.

Achieving Conceptual Understanding

Having looked at what conceptual understanding is, our next question is, how do instructors help students to achieve it? More specifically in our case, how does the designer of a video game help students achieve it? This, as we shall see, is not an easy matter. Nor is it easy to ascertain whether or not a student has achieved it—even in face-to-face contact, let alone through game artificial intelligence.

The accepted wisdom for introducing a new concept in a fashion that facilitates understanding is to begin with several examples. For instance, the celebrated American mathematician R. P. Boas had the following to say on the issue, in an article entitled "Can We Make Mathematics Intelligible?" which was published in the *American Mathematical Monthly*, Volume 88 (1981), pp. 727–731:

Suppose that you want to teach the 'cat' concept to a very young child. Do you explain that a cat is a relatively small, primarily carnivorous mammal with retractable claws, a distinctive sonic output, etc.? I'll bet not. You probably show the kid a lot of different cats, saying 'kitty' each time, until it gets the idea. To put it more generally, generalizations are best made by abstraction from experience.

This idea is appealing, but not without its difficulties, the primary one being that the learner may end up with a concept different from the one the instructor intended. The difficulty arises because an abstract mathematical concept generally has fundamental features different from some or even all of the examples the learner encounters, which is, after all, one of the goals of abstraction!

An important illustration of this issue that has been studied extensively is the modern mathematical concept of a *function*. For instance, the Israeli mathematician and mathematics educationalist Uri Leron, in his article *Mathematical Thinking and Human Nature: Consonance and Conflict,*[3] wrote:

> According to the algebraic image of functions, an operation is *acting on an object.* The agent performing the operation takes an object and does something to it. For example, a child playing with a toy may move it, squeeze it, or color it. The object before the action is the input and the object after the action is the output. The operation is thus transforming the input into the output. The proposed origin of the algebraic image of functions is the child's experience of acting on objects in the physical world. . . . Inherent to this image is the experience that an operation *changes its input*—after all, that's why we engage it in the first place: you move something to change its place, squeeze it to change its shape, color it to change its look.
>
> But this is not what happens in modern mathematics or in functional programming. In the modern formalism of functions, nothing really changes! The function is a "mapping between two fixed sets" or even, in its most extreme form, a set of ordered pairs. As is the universal trend in modern mathematics, an algebraic formalism has been adopted that completely suppresses the images of process, time, and change.

Leron and others have carried out several studies demonstrating that many mathematics and computer science students at universities have formed erroneous concepts of functions; for instance, often assuming that applying a function to an argument changes the argument. Such is the power of the original examples, that

---

[3] Proceedings of the 28th Conference of the International Group for the Psychology of Mathematics Education, Vol. 3, 2004, pp. 217–224.

even when presented with the correct formal definition of the general, abstract concept, the learners assume features suggested by the examples that are not part of the abstract concept. It can take an experienced instructor some time to uncover such a misconception.

Given current capabilities in artificial intelligence, we cannot assume that a software system will be able to do this either. Of course, you could design a system that checks for any specific misconception you anticipate in advance and that responds accordingly—something video games can provide—but there is no limit to the number or subtlety of errors that a learner might make. Thus, while conceptual understanding is a goal that educators should definitely strive for, we need to accept that it cannot be guaranteed, and accordingly we should allow for the learner to make progress without fully understanding the concepts.

NRCR seems to accept this problem. Rather than insist on full understanding of the concepts, the committee summarized what they meant by "conceptual understanding" this way (p. 141), "... conceptual understanding refers to an integrated and functional grasp of the mathematical ideas." I believe the key term here is "integrated and functional grasp." This suggests an acceptance that a realistic goal is that the learner has sufficient understanding to work intelligently and productively with the concept and to continue to make progress, while allowing for future refinement or even correction of the learner's understanding in the light of further experience. It is possible I am reading something into the NRCR's words that the committee did not intend, in which case I suggest that in the light of further considerations I am refining the NRCR's concept of conceptual understanding! It is this relaxed notion of conceptual understanding that I am referring to by my term *functional understanding*. It means, roughly speaking, understanding a concept sufficiently well to get by for the present.

For the remainder of this book, whenever I use the term "conceptual understanding," you can read that as "functional understanding." I'll continue to use the more common former term, though as I have indicated, I think it is almost always used to mean what I am calling functional understanding. Indeed, I don't see how, in practical educational terms, it can mean anything else. Occasionally, where I think the distinction is important, I'll explain. Since the distinction I am making is somewhat subtle, let me provide a dramatic example. Since Newton was the person who invented calculus, it would clearly be absurd to say that he did not understand what he was doing. Nevertheless, he did not have a conceptual understanding of the concepts that underlie calculus as we do today—for the simple reason that those concepts were not fully worked out until late in the nineteenth century. Newton's understanding, which was surely profound, would be one of functional understanding. Euler demonstrated similar functional understanding of infinite sums, though as with Newton the concepts that underpin his work were not developed until much later.

One of the principal reasons why university mathematics majors progress far more slowly in learning new mathematical techniques than do their colleagues in physics and engineering is that the mathematics faculty seek to achieve full conceptual understanding in mathematics majors, but what future physicists and engineers need is functional understanding. Teaching with functional understanding as a goal carries the responsibility of leaving open the possibility of future refinement or revision of the learner's concept as he or she progresses, so the instructor should have a good grasp of the concept as mathematicians understand and use it.

How Well Are Students Doing at Acquiring Functional Understanding?

Defined in terms of performance, the more operational-oriented notion of conceptual understanding that I am calling functional understanding (and which I believe is what the NRCR were actually thinking of) is something that can be assessed by measuring performance. So how well are United States schoolchildren doing on this score? The answer is not well at all.

About 90% of American 13-year-olds can add and subtract multi-digit numbers, but only 60% can construct a number given its digits and their place values (e.g., in what number does 3 represent three units and 5 represent five tens?). In general, more students can calculate successfully with numbers than can work with the properties of the same numbers (NRCR p. 136). Similarly for fractions, only 35% of 13-year-olds can correctly order three fractions, all in reduced form, and only 35%, when asked to select from a given collection the one that lies between 0.03 and 0.04, chose the correct response (NRCR p. 137). It is of course possible that these same 13-year-olds can and will function as most adults in the world do without this level of knowledge—just as the Brazilian children did. So it is possible that they have a functional understanding at some level not measured by the test questions. In any event, part of the reason for their lack of functional understanding may be lack of desire. As every teacher discovers early in his or her career, many students, many parents of students, and many adults who set out to learn some mathematics (perhaps a majority of individuals in all three categories) do not seek to *understand* mathematics conceptually to any variety or depth. Rather, their goal is simply to be able to *do* it reasonably well when the need arises, without understanding what they are doing; that is, their goal is solely procedural fluency.

A designer building a game intended to help people achieve mathematical proficiency in the sense of NRCR faces the questions: to what extent is functional (conceptual) understanding (i) helpful and (ii) necessary in order to achieve an acceptable and lasting procedural fluency? There is little doubt regarding the first question. Evidence indicates that understanding helps performance. The practical question that this raises for the video game developer is this: since conceptual understanding (either the functional or complete variety) takes time and effort to acquire, how much time and how much effort should you devote to *explicitly* helping the student achieve conceptual understanding? And when you have answered that question, a related question is: since relatively little is known about how conceptual understanding is achieved and how teachers can help students acquire it, how are you—acting at arm's length—going to help them progress toward this goal? I return to this question later.

It is less clear the degree to which question (ii) has a positive answer. Indeed, there is some evidence that suggests that achieving (initially without understanding) a degree of procedural fluency is a necessary step on the path to eventual conceptual understanding. What is known, however, is that, if the acquisition of procedural skills is not eventually accompanied by conceptual (or functional) understanding, then the resulting skill is both brittle and subject to deterioration over time. In that respect, conceptual understanding is an extremely important goal. The strongest factor going in our favor in this respect is that all learning that takes place in a video game does so in an appropriate context of use, a feature whose importance I have already noted.

Procedural Fluency

Procedural fluency refers to skill in performing procedures flexibly, accurately, and efficiently, and to knowing when and how to use those procedures. Here is how NRCR describes it:

> Acquiring proficiency takes time. . . . Students need enough time to engage in activities around a specific mathematical topic if they are to become proficient with it. When they are provided with only one or two examples to illustrate why a procedure works or what a concept means and then move on to practice in carrying out the procedure or identifying the concept, they may easily fail to learn. To become proficient, they need to spend sustained periods of time doing mathematics—solving problems, reasoning, developing understanding, practicing skills—and building connections between their previous knowledge and new knowledge. (p. 135)

It is in this area that I feel that video games have the potential to make a huge impact. There is hardly any limitation to the extent to which well-designed video games can provide exactly what the NRCR suggests. The degree to which that leads to greater understanding remains to be seen, once we have enough well-designed

mathematics education video games. The design of those games will have to involve experienced and knowledgeable mathematics educators.

Strategic Competence

Strategic competence refers to the ability to formulate, represent, and solve mathematical problems that arise in some real-world or mathematical domain. This
is an ability of enormous significance for a country like the Unites States whose
economy depends upon constant innovation in science and technology, yet we
perform extremely poorly when compared with many of our competitor nations.
The National Assessment of Educational Progress (NAEP), begun in 1969, is a regular assessment of students' knowledge and skills in school subjects. It includes
a large and representative sample of United States students of about 9, 13, and
17 years old. For more than 30 years, the NAEP educational assessment reports
have demonstrated consistently that one of the greatest deficits in United States
students' learning of mathematics is in their ability to solve word problems.

NRCR based its conclusions on the 1996 assessment, but performance levels
have been fairly static since the first NAEP report came out. In the 1969 assessment, when asked to add or subtract two- and three-digit numbers, 73% of fourth
graders and 86% of eighth graders gave correct answers. But on a multi-step addition and subtraction word problem involving similar numbers, only 33% of fourth
graders gave a correct answer and 76% of eighth graders. (NRCR p. 138) On the
23 problem-solving tasks given as part of the assessment in which students were
asked to construct an extended response, fewer than 10% responded satisfacto-
rily on about half the tasks, and the rate of satisfactory responses was greater than
25% on only two tasks. (NRCR p. 138)

An aside is in order here. It is likely that these very same kids, when placed
in a situation like the Brazilians where the mathematics mattered, may have been
able to do exactly what was not being measured on the NAEP tests. The real question here is whether the tests of the types talked about both here and in the Brazilian studies measure the understanding of a concept. Surely, the true test is not the
paper-and-pencil type with which the United States has become overwhelmed,
but the "doing" type test that the Brazilian kids passed with flying colors. It is of
course simply too expensive to do this type of testing on a wide scale so we resort
to the paper-and-pencil variety and complain that the kids know little or nothing.
Video games have a clear potential to be used as an assessment tool for everyday
basic mathematics.

Returning to my main theme, performance on word problems declines dramatically when additional features are included, such as more than one step or
extraneous information. Small changes in problem wording, context, or presenta-
tion can also yield dramatic changes in students' success (NRCR p. 138). In fact,

studies have shown that word problems can be detrimental to a student. They are always artificial, have little true meaning, and are most often assumed to be "real world" when in fact they are far from it. Since United States children perform worse than their counterparts in other countries, much of the problem must lie in the teaching, though there is no denying that students the world over find word problems hard and always have. In any case, if there is a better way to achieve the same educational goal, we should adopt it. And there is.

As the study of the Recife schoolchildren demonstrated, ordinary people can do extremely well at solving often difficult, multi-step math problems when the problems arise in the real world. I have already suggested that much of the learning difficulty may be that the (often very brief) statement of the word problem is not adequate to put the learner into a cognitive context sufficiently like the real world situation the problem is ostensibly about. When a player engaged in a well-designed math ed video game in a virtual world encounters a mathematics problem, he or she will already be in the world in which the problem arises. This is one reason why it is crucial that the mathematical tasks the player has to perform arise naturally in the game. If this happens, they will be in a situation much more like the Recife stallholders than of a student sitting in a classroom reading a word problem in a math textbook. As a result, math ed video games in virtual worlds are ideal for developing strategic competence and assessing understanding and learning.

Adaptive Reasoning

Recall that adaptive reasoning is the capacity for logical thought, reflection, explanation, and justification. Assessing adaptive reasoning is most easily done in conjunction with other skills, often conceptual (or functional) understanding. For example, in one case cited by NRCR, students were asked the following question:[4]

If 49 + 83 = 132 is true, which of the following is true?
49 = 83 + 132
49 + 132 = 83
132 − 49 = 83
83 − 132 = 49

Only 61% of 13-year-olds chose the right answer, a figure far lower than the number who could successfully carry out the computation given. Another question asked:

Estimate 12/13 + 7/8. Is it 1, 2, 19, or 21?

---

[4] See NRCR p. 139 for details of the examples given in this section.

Here, 55% of 13-year-olds chose 19 or 21 (the totals of the numerators and denominators, respectively) as their answer. It is of course easier to recognize that the two fractions are each slightly less than 1 and conclude that their sum will be just under 2 rather than to perform any kind of computation. Nevertheless, the students apparently felt compelled to compute something, rather than to reason about the sizes of the fractions given.

Experience in mathematics teaching since the time of Euclid indicates that adaptive reasoning is difficult to achieve, for the reason suggested above. Learning mathematics is a continual process of learning new techniques. This process requires so much effort that it is easy for a student to get lost in the trees (or the undergrowth) and fail to see the forest, or even be aware that there is a forest. All of their focus is on learning how to perform the most recent operation and that becomes their conception of what mathematics is. As a result, they feel "compelled to compute something."

I believe that video games that include quests are well suited to developing adaptive reasoning.

Productive Disposition

NRCR describes productive disposition as "a habitual inclination to see mathematics as sensible, useful, and worthwhile, combined with a confidence in one's own ability to master the material." Students' disposition toward mathematics is a major factor in determining their educational success. Students who view their mathematical ability as fixed, and test questions as measuring their ability rather than providing opportunities to learn, are likely to avoid challenging problems and be easily discouraged by failure (NRCR p. 131).

If they are properly designed, video games have enormous potential to develop productive disposition in learners. Because the mathematics problems the students face in the game arise naturally, the player's very experience is one in which doing math is useful *in the game*. If the game world is sufficiently like the real world, I expect this to spill over automatically to a sense of mathematics being useful in the real world as well. Future studies will be able to test this. Indeed, there is evidence that players are likely to be well disposed to reaching this conclusion before they ever start to play. In the 1996 NAEP, 69% of fourth graders and 70% of eighth graders agreed that mathematics is useful for solving everyday problems.

I should stress again that a video game can teach mathematics to the students only if there has been a well thought out plan for which mathematics arises naturally in the game. When a game action requires knowledge of mathematics, you have to consider the learning progression. Early tasks or quests have to be easier ones, or players might quit in disgust. For an educational video game to succeed, the placement of the mathematical tasks may be one of the more important aspects.

 # Building a Successful Math Ed Video Game

To some readers the chapter title might suggest a how-to manual for game developers, but that is not my intention. Rather, my primary audience remains mathematics educators who either want to adjust their pedagogy to facilitate the incorporation of video games into the mathematics learning they provide or perhaps want to develop a video game themselves, either on their own or in collaboration with an experienced game designer. My words can perhaps benefit the game designer by describing the pedagogic principles a game designer—or a mathematics educator working with a game designer—might consider in order for the resulting game to yield successful learning outcomes in terms of conceptual understanding and mathematical thinking. Of course, pedagogic principles for video game design can only be effective if they are actually implemented. So, I will make some specific implementation suggestions. They are, however, merely illustrative examples to get such a reader started. Experienced game designers will likely know everything I say about principles of game design, but most mathematics educators—my primary audience—probably do not.

As with any educational enterprise, the first step is to try to understand the learner's situation. One obvious difference between math textbooks and video games is that most students need to be pressured into obtaining and reading a math textbook, but over 97% of students buy and play video games because they want to. Of course, those same students might spend many hours voluntarily reading novels, so it's not just a question of the medium; the message matters too. My point is that video games are a different medium from print, so we should start by assuming all bets are off.

Why Do People Play Video Games?

What persuades an individual to purchase a video game, and then spend many hours—sometimes hundreds of hours—playing it? It's easy to dismiss it as "mere entertainment," but what is it about the video game experience that makes it entertaining? It's definitely not the same kind of entertainment as television. Most television is essentially passive, while video games are interactive and highly engaging. Many of us find ourselves falling asleep in front of the television set, but with video games the opposite is the problem—dragging yourself away to go to bed can require considerable willpower! Moreover, most video games are hard and require considerable skill development.

Video games succeed because they tap into some very basic aspects of human cognition. One is that humans evolved to act smart. I'm not talking about "being smart" in the sense of someone with a college degree or a high IQ; rather, acting smart means acting in a way that distinguishes a person from, say, a sheep. Acting smart is not something we have to learn; it's a basic instinct. We acquire mastery of new physical skills, we make plans, we learn from our experiences, we anticipate future events, we solve problems, we take steps to avoid danger, and we collaborate when necessary.

When we are successful in acting smart, and achieve one goal or another, we are pleased and take pride in our accomplishment—pride in our accomplishments being another product of evolution. Some of us are lucky enough to have jobs that involve frequent opportunities to gain satisfaction from acting smart; for many people the enjoyable acting-smart experiences are largely restricted to their leisure activities.

Notice the word "act" in the previous paragraph. What humans do is *act* smart. "Being smart" means something altogether different. Some of us gain enjoyment from simply knowing things—knowledge for its own sake—but I suspect we are a minority. We all get kicks from *acting* smart, however. Acting smart is what evolution equipped us for, along with the reward system (pride, satisfaction, survival) for acting smart with success. (Recall the earlier discussion of the distinction between knowing *that* and knowing *how*, in Chapter 4.)

Evolution also equipped us with something else of relevance to the video game designer: we are social creatures. Most of us seek out groups to join, and even the "loners" among us generally identify with certain kinds of people, even if they rarely if ever seek to spend any time with them. We are pleased by recognition and approval from our chosen group. This instinct to identify with a particular group—to be or feel like a member of a clan—is a powerful one. It is particularly evident with political parties, sports teams, and religious organizations. I doubt if there is a single person who agrees with everything the Republican party stands for, yet many Republican voters will not voice objection to the party line, and likewise for the Democrats. Sports supporters sometimes engage in fights with supporters of rival teams. And millions of humans have been, and continue to be, victimized in the name of one religion or another. Observations such as these are generally made by way of lamenting the shortcomings of the human character—and perhaps they are—but their very universality provides us with a powerful message when it comes to education.

It's Actually More about Being than Doing

An instinct to act smart and to learn new ways to act smart, identifying with and being a member of a group, and gaining pleasure from both are all powerful features of people that the developer of a video game can tap into in order to design a compelling game experience—which is precisely what they do.

Mathematics is a particular way of thinking; everyday mathematics is a particular way of thinking about the world. The central goal of everyday math ed is to develop the ability to think about the world in that way. I suspect everyone would agree with this. Where many people go wrong is failing to follow through on what this really entails. Namely, thinking about the world in a mathematical way involves assuming a particular identity—this was our discussion of semiotic domains in Chapter 5. In order to do mathematics with any success, you have to *be a mathematician*, or at least *mathematically able*. This does not mean being a professional mathematician, or being as skilled as a professional mathematician, or even being on a par with a good mathematician. I'm not restricting my attention to being good at math. Rather, in order to be competent in basic math, you have to *be* a mathematician or a mathematically able person in the sense of assuming the identity of someone who thinks mathematically and is aware of that capacity.

Mathematics education, when it is successful, is only partially about people learning how to *do* mathematics; it is also about helping them adopt a particular identity—that of *being* at least a mathematically able person, and possibly even a mathematician. Of course, being a mathematician or a mathematically able person entails the ability to do math and can be achieved only by doing math. But if the learner's and teacher's focus is solely on "doing math," the educational process will likely fail to meet its goal. If, however, the learner is able to acquire the identity of "being a mathematician" or "being mathematically able," then everyday mathematics automatically makes sense and becomes doable. Not necessarily easy, but doable.

It is regrettable, I think, that the NRC's *Adding it Up* did not highlight this key feature of mathematics education. The closest they came was the inclusion of productive disposition as one of their five interwoven strands, defining it as "a habitual inclination to see mathematics as sensible, useful, and worthwhile, combined with a confidence in one's own ability to master the material." I suspect the reason why *Adding it Up* did not explicitly discuss identity is that it is only since the report came out in 2001 that educators have begun to take seriously the issue of self-acknowledged identity in education. In part, that recognition has come as a result of studying the success and effectiveness of video games in getting people to master masses of often difficult and challenging material and skills.

The problem facing the video game designer is this: to be successful, the game has to continually present the player with new challenges. The player has to learn not only what the game is about, but also *how to play* the game, and moreover has to do so without any aid other than what is supplied within the game itself. For most games, the initial learning curve is extremely steep. Indeed, many older people who start to play—often in order to play with a child or grandchild—find it frustratingly difficult and in many cases give up, saying they cannot make any sense of what is going on and what they are supposed to do. This is, of course, essentially the same reaction that mathematics teachers see every year in the math class. So what makes video game players (at least the younger ones) persevere until they have gotten over that steep initial learning curve and mastered the basic moves of the game, while students drop out of math classes in droves?

A large part of the answer is identity. In many games, the very first thing the players do is select their in-game identities—their avatars. From then on, everything that the game designer presents to the players is directed not to the players themselves but to their avatar identities. Players who choose to be warriors, for example, are presented with tasks and challenges designed to make their player-avatars better warriors. Players who choose to be priests will be thrust immediately into activities designed to make them better priests. And so on. In other words, a player is not being asked to "learn how to do X." Rather, it is all about "becoming a (better) X-er."

A key feature of this process is that the players make the initial choice of the identities they will assume in the game. This choice often involves a great many specific features that allow the players to pick characters that they feel comfortable with and reflect characteristics they either have themselves, or would like to have, or simply wish to adopt. There is something of an illusion about this process, since the practicalities of game design mean that once the game starts, the number of actual options available is relatively small. Yet it works. Players quickly identify with their characters and, for all the natural reasons that evolution has equipped us with, want to make their characters better able to succeed. But can we adopt the same strategy to teach mathematics? What happens if the character is a mathematician, or at least a character who is good at math, or at the very least someone who *can do* some math? Can we use the identity trick to put people in a position to want to learn math?

In real life, a small number of people do come to see themselves as mathematicians. I was one of them. For whatever reason, around 16 years of age, "mathematician" was an identity I began to recognize in myself. It seems to me unlikely that there is any hope of getting more than a small minority to see themselves as mathematicians, either in a real-life setting or in a game world. (I happen to think it is not desirable either; life thrives on diversity.) But if the goal is to be "mathematically able," then I see no reason why this cannot be achieved. Interestingly, more Americans did start to think of themselves as "mathematically able" during the 1960s, when the United States put a national priority on mathematics, science, and engineering in its struggle with the Soviet Union. So there is room for leverage with this approach. But the very word "mathematics" conjures up such a powerful—and for most people negative—image in today's society that asking people (particularly young people) to assume that identity up front seems a hopeless task. A more realistic goal is that after players have completed enough of a game, (1) they will be mathematically able, (2) they will recognize that fact, and (3) they may even feel comfortable with the term "mathematically able" being applied to them.

Of these three goals, you may have to settle for only the first two. In today's society, expecting adolescents to be comfortable with the term "good at math" may be asking too much. Remember, thinking of yourself as a mathematically able person does not entail being good at it, any more than a group of children kicking a ball around on a piece of wasteland and seeing themselves as soccer players means they are professionals. It's more an attitude of mind than mastery or skill.

The fact is that many successful games already on the market actually do require that players carry out certain kinds of mathematical thinking, among them logical problem solving, comparison of numerical ratios, scale conversions, etc. They just don't draw attention to the fact that it is mathematical thinking. But the designer of a math ed game has to go a significant step further, since the players should come to adopt the identity of "being a mathematically able person" (i.e., being someone who can think mathematically) not only in the game but subsequently in the real world.

The best way to achieve this, in my view, is to build the entire game world and game around key mathematical learning experiences—that is to say, to design the game world and construct the game so that it is perfectly natural for player-characters to think mathematically. The intention is that when a player tries to advance his or her character—to make the character a better warrior, priest, etc.—this will entail thinking mathematically about naturally arising aspects of the game world and game play. Thinking mathematically should simply be part of what that character does in that world. The mathematics should not be hidden; the players should know they are doing math! But that math should arise naturally in the game, it should have meaning in the game, and it should make sense in the game. If the game design is successful, players will do the math as part of the complex identities of the character that represent them in the game. Incidentally, this is very different from the approach that some have suggested of taking a video game whose shelf life has long since expired and embedding some mathematical tasks into it. That might be much cheaper than building a game from scratch, and it might work. But for a variety of reasons, the odds seem heavily stacked against it. In any case, it's not the approach I am suggesting.

Five Key Principles for Building a Mathematical In-Game Identity

If all that were required to get people to learn mathematics were to get them to "see themselves as mathematically able," we could make today's high level of mathematical illiteracy vanish in a generation. There is, after all, no shortage of individuals who see themselves as writers or poets or musicians who have never mastered the basics of those crafts. I began this chapter with a discussion of mathematical identity not because it is the only issue—it most certainly is not—rather because it is a crucial factor that is often overlooked. But a large part of assuming that identity of "being a mathematically able person" is learning how to think mathematically—to "do math" if you like—although that phrase "do math" is all too frequently taken to mean mindlessly manipulating symbols, without the full engagement that comes with genuine mathematical thinking.

As I have lamented earlier, rule-following symbol manipulation without engagement or understanding is often assumed to be what "doing math" is all about, an erroneous perception that is hardly surprising given that our current educational system presents mathematics that way to most students. When it comes to helping people learn how to think mathematically, those of us in mathematics education can learn from the video game industry, even if our goal is mathematics education in a traditional setting, not a video game. Commercially successful video games follow a well-established path to getting their players to learn what is required in order to play the game. The principles employed are known to knowledgeable educators, but they are rarely used in schools. Indeed, the traditional educational curriculum is set up in such a way that makes it all but impossible to proceed in the way I am about to describe.

Five features of good game design are particularly crucial in order for the game to develop mathematical thinking in players who come to the game believing that they cannot do math. All but one of these features are conspicuously absent in most traditional mathematics classrooms.

Learning by doing. The first feature is that the player is never put in a position of having to "learn something" prior to playing the game in order to play the game. Rather, learning is achieved by playing the game. Everything the player is presented with is part of the game and has meaning within the game world. The player being in the game world—and it definitely is a world—is something that is not available in the schoolroom. Thus, while I believe there is much that the traditional mathematics education system can—and should—learn from the way video games are constructed, it may be that video games simply have some significant advantages that cannot be matched in the classroom. Moreover, the player can always progress in the game, and be aware of progressing.

In many games, this initial learning-by-playing is arranged by having the "newbie" or "noob" (gamer parlance for the beginning player) find himself or herself in a fairly safe, restricted part of the game world, where there are no major threats and where things happen slowly—usually at the player's own pace. This allows the player to master some key features of the interface and learn how to manipulate the avatar and use it to do things in the environment.

There is usually a large amount of learning that takes place during this newbie phase, but it is all achieved through the player's action in the game world, and everything that the player does is under his or her direction (or at least the player thinks that is the case). Incidentally, my use of the word "action" should not be taken to imply that everything is achieved by the player moving his or her avatar around. For some games there is a significant amount of reading involved. But that reading is always directed toward the player's game identity (and not the player), it is oriented to character development, and it is transparently aimed at improving the avatar's in-game action. It is often about game lore or how that player's character is supposed to behave.

Self-paced learning. Once the player has learned the basics, he or she is encouraged to progress further. Note that word "encouraged." The player is not forced to do anything. The second key feature of game design that is critical for mathematics learning is to allow players to proceed at their own pace. Slow or hesitant players can take as much time as they want before moving forward. Bold players can move forward as soon as they want. If they move too far ahead of their acquired skill level, something "bad" will happen to them in the game (usually game death, which as I noted earlier is never permanent, and usually results in a cost primarily of the player's time). They will then be forced to step back and make sure they have the skill they need to progress—a task for which they are now highly motivated!

Exploration. Because players proceed at their own pace (at least during learning phases), they can take time to explore, try things out, and become familiar with new ideas and skills. Most mathematics teachers are well aware of the importance of self-pacing and giving students sufficient opportunity to explore and try things out, but the current school system, with its packed curriculum and the emphasis on "assessment," does not allow sufficient time for either. (I put the word "assessment" in quotes because all that is actually assessed is the student's ability to perform in the assessment tests.)

Immediate use. In a video game, when a player learns a new fact or skill it is put to immediate use. The game designer makes new information or the learning of a new skill available only when the player needs it in order to progress in the game. Moreover, this is done in such a way that at that stage of the game the player can (if they think about it) see the immediate need for that new information or skill. This is often done so skillfully that players do not realize it is happening. The game designer sets up game situations so that new facts and new skills are acquired on demand, and just in time to be used.

Regular tests. Finally, video games present the players with frequent tests to see how well they have mastered the latest facts or skills. What's that you say? Tests of what has been learned? In video games? You bet there are. In fact, the entire progression through a typical video game consists of a repeating cycle of learning and testing, learning and testing, etc. Everything that is learned is tested immediately and repeatedly, whether it is a manipulative skill such as making your character perform a particular action, perhaps a combat move or a jump up to a high ledge, or planning out and executing an intricate series of moves through a large group of threatening mobs. (Mobs are the bad guys—NPCs intent on killing you.) In fact, it's the *enjoyment* of taking and passing the "tests," often after several failures, that *motivates* players to learn. Though a few of us always found school math tests fun (because we could do well at them!), most people do not, so this phenomenon in video games is a total reversal of what happens at school. Obvious factors at play in a video game (but not the school classroom) that may help explain this phenomenon include

- the player has lots of time to prepare for an upcoming "test" at his or her own pace,
- the learning and the test are carefully designed, and fine-tuned through many cycles of user testing, for the test to be doable yet at the limit of the player's current ability in order to give satisfaction on completion,
- the entire game is constructed so that the player comes to realize that nothing is beyond his or her capabilities, provided they work at it,
- failure is not public, and there is no humiliation for failing, and
- if the game is well designed, failing at a task carries with it the belief "I have not yet succeeded" (so it is more a challenge to keep going than a discouraging blow).

Building a Math Ed Video Game: First Steps

As I noted in Chapter 3, there are many different kinds of video games, and the first step facing the math-ed video game designer is to decide which kind to choose. Money, time, game design expertise, game development skill, and other factors will all play a role. In making the choice, the designer needs to be aware of the different features of video games that can be utilized to provide mathematical learning experiences. The following list of video game features is an amalgam of a number of such lists generated by game developers and researchers who have studied video games:

1. assets,
2. communication,
3. feedback,
4. marketplace,
5. narrative,
6. ranks and levels,
7. reputations,
8. rules,
9. self representation,
10. tasks,
11. teams,
12. time pressure,
13. worlds.

Some games have only some of these features. MMOs have them all—though they are by far the most difficult and expensive to build. The designer has to work with whatever features the particular game genre has to offer. Though lists such as this are generally drawn up as general principles of game design, I'll look at each feature in turn with regard to opportunities to facilitate learning of basic mathematics.

Assets. A player's character has a number of different assets, including character features (maximum health, strength, intelligence, stamina, speed, etc.), powers and skills (spell casting ability, fighting skills, first aid ability, healing powers, etc.), armor, weapons, ammunition, pets, mounts, money, potions, food, and so on. Game play provides many opportunities to acquire new assets and to strengthen or otherwise improve existing ones. Players spend considerable time, effort, and game money improving their inventory of assets. The acquisition of assets thus provides players with a major incentive to play and to take on new challenges.

A game designer can use the desire for particularly cool and powerful assets to motivate players to engage in tasks that require mathematics. Where possible and appropriate, the designer can also embed mathematics in the choice, acquisition, manufacture, and use of various assets.

A lot of mathematics pertaining to game assets is not so much action by the avatar in the game world, but rather action by the player at the level of the game interface. This kind of activity is called game management. Particular examples that offer scope for using mathematics are status bars (e.g., health, manna), the player's inventory (e.g., loot, potions, special tools), and the character's profile (e.g., weapon strength, armor rating, reputation).

For instance, building a powerful character and achieving success in combat or other activities generally involves making wise choices that depend on numerical comparisons—such as comparisons with alternative choices of equipment or with the profiles of imminent enemies or terrain obstacles. In my case as a *World of Warcraft* player, I found that optimal allocation of talent points often involved sophisticated reasoning involving cumulative percentages.

Although game management activities are an integral part of playing the game, they are not *in* the game itself. Thus, when a designer builds a game interface to require more mathematics than is usual in comparable video games, care needs to be taken to ensure that game management activities do not detract from, or even destroy, the overall game experience.

Communication. In many current multiplayer games, part of the computer screen is devoted to a chat window, which players can use to communicate with one another by text. This allows players to socialize and cooperate with others. Some of the more challenging tasks in the game can be completed only when players cooperate in a team, and for those tasks the communication channel is crucial. (Expert players often enhance their play by adding an additional voice channel; some game manufacturers provide a voice over internet protocol (VoIP) channel as an optional part of the regular game interface.)

In games intended for adults (such as *World of Warcraft*), the only built-in restriction on what players can type into the communication channel is the automatic filtering of obscene or otherwise offensive words. For a game designed for younger players, there are legal restrictions that have to be followed. For instance, it would be important to restrict players to chat only with players on a specified list, agreed to in advance by the teacher or the parents. Because the social setting influences learning, and given the recognized benefits of collaborative learning, player-to-player communication clearly offers considerable potential to enhance the learning process.

Feedback. I discussed earlier the importance in education of appropriate and timely feedback (see, in particular, Chapters 3 and 8). By their very nature, video games are able to provide this feedback in a big way—and they do so. Much of the compelling nature of video game play stems from a game's ability to provide constant, instant, and personalized feedback that is directly relevant to what the player has just done and wants to do next. That benefit excites those of us who have became interested in using video games in mathematics education.

Marketplace. Another obvious mathematically related game element that an educational game could share with *World of Warcraft* and other games is having a game economy—banking, buying items from and selling items to NPCs, trading goods with other players, and buying and selling in the auction house. A natural choice the designer could make for a game geared toward math ed is to make the various financial transactions explicit. But there are problems with that approach. Let me explain.

Game management occurs outside the game play itself. It should not interrupt the flow of the play. In-game financial transactions are, by their nature, part of the game play. A crucial element of a successful game is pace. If any one action slows down the game too much, many players are likely either to avoid those actions or, if that proves impossible, abandon the game altogether. Even in a hugely successful game like *World of Warcraft*, financial transactions already slow the pace, despite the fact that all of the math is done by the game's artificial intelligence, or AI. (All the player does is click on the *Buy*, *Sell*, or *Trade* button in an interface window.) There is a risk in losing player involvement if functions that would slow the game pace are introduced at such a stage.

Another challenge in making the players do the financial arithmetic is that one of the main incentives to keep playing is the acquisition of in-game money. In *WoW* this takes the form of copper, silver, and gold coins, with 100 copper coins being equivalent to one silver coin and 100 silver coins equivalent to one gold coin. Beginning players start out with a small number of copper coins, and acquire more for completing quests in the game and occasionally by other means. The more they play the game, the more opportunities they meet to acquire larger sums of money. Once a player is at an experience level that she or he starts acquiring silver coins, individual copper coins become irrelevant. Thus, any hope the game designer had to use game money to provide the player with practice at handling decimals by having to deal with monetary amounts like 15s 27c (15 silver and 27 copper) will be in vain, as the player simply ignores the copper and views the amount as 15 silver. Indeed, once players are in the silver range, they typically don't even bother to pick up any copper coins they come across. The same thing happens again when they get into the gold coin range; silver coins become ignored and irrelevant. In the real world, there are people who still exercise care in keeping track of the pennies, but in video games that simply is not the case.

On the other hand, the ingenious game designer could introduce a mechanism to force the use of mental arithmetic. In fantasy-world video game, for example, one way would be to introduce a town that uses its own currency, forcing players to go to designated NPCs to exchange currency in order to buy or sell in the town, and script the interactions with those NPCs so that the player has to do the math correctly to complete the transaction. Something that could otherwise become a distracting chore if it were a regular feature of all transactions can work if it arises only in a particular circumstance, and where it does so naturally. But engaging in such transactions should be fairly infrequent and mostly left to the player's choice. It should be possible to avoid them, even if at the cost of not acquiring some particularly cool or useful items. This example illustrates a point I made earlier that one of the beauties of fantasy worlds from an educational standpoint is that the designer gets to invent the world, and there is almost always a way to engineer a desired activity. The trick is to ensure that activity is elective and make the reward for success sufficiently attractive.

Narrative. It would be easy to dismiss this as nonessential to the learning component of an educationally oriented game, but that would be a major mistake. People familiar with the video game world long ago recognized that the back story is crucial to the success of a game. Game projects that began with the software development and left the development of the game narrative until later invariably failed—the games simply did not sell. When an author sets out to write a novel, he or she begins with a story idea and the focus is on developing that idea. The task of expressing the story in words on paper is secondary; it's merely the delivery method. Of course, the quality of the writing is important. But as a glance at the bestsellers lists will confirm, good stories can sell even if the writing is poor, while no amount of brilliant writing can make a success of a poor story. The same is true for video games, but because the development of the delivery mechanism is so labor intensive, expensive, and high tech, it is easy to overlook the fact that the game mechanics are simply the equivalent of the novelist putting his or her story into words on paper.

Ranks and levels. Much of the impetus for players to keep playing comes from watching and experiencing their characters gain experience points (XP) and advance upwards through the various game levels and character ranks.[1] Level design is a crucial part of game design that requires considerable care, and online-

---

[1] It's not just the acquisition of the higher rank that drives game play. At least as important, and most likely considerably more significant, is the social status that higher rank confers among the player's friends.

game developers constantly monitor game play and adjust the levels to ensure that players have the best possible game experience.

Reputations. Formalized (quantitative) reputation systems are an important part of competitive (player versus player) play, and for players who like such games, reputation is as big a driver as leveling. For a mathematics educational game, competitive play with a reputation system may not be appropriate, however, as it could be a strong disincentive to mathematically weaker students. This does not mean that reputation is not important. We all like others to see our successes. But it may be better to leave it to the players to acquire reputation informally among their own communities as they see one another in action, rather than through a public leaderboard.

If a designer does want to have competitive play with a leaderboard—and there is no doubt it appeals to many players/students—attention should be given to how the rankings are displayed. For instance, displaying the scores of the very best players can be discouraging to less experienced, or innately less skillful, players. On the other hand, presenting a player's score in a list with the five next better scores above and the five adjacent lower scores below can provide a powerful stimulus to try to move up one or two rankings and to avoid being caught from below. Even then, it's probably a good idea to wipe the board clean once a week, to avoid higher-ranked players resting on their laurels.

It is also worth considering using game AI to engineer public recognition for players who don't excel in the most obvious activities. Everyone likes to be recognized, and there are few people, if any, that don't do well at something. The overriding goal, after all, is education. That means taking people and making them more knowledgeable or better at certain things, not trying to identify the next Nobel Prize winner.

Rules. It is of course the presence of rules that determine a game. Video gamers as a community are known to enjoy looking for ways to break or circumvent rules, for example by means of "cheat sheets" posted on websites. Game designers know this and design their games accordingly. A designer should expect that a great deal of the learning that will (we all hope) accompany the playing of the game will occur "outside the rules" in that way, with players learning from one another.

One feature you find in many MMOs is that players can buy or sell desirable game items outside the game, on eBay, or through other online sites. Most educators probably feel this is inappropriate for a children's educational game, though it would surely help develop everyday math skills (assuming the child had access to a credit card). But what about allowing players to trade or sell (for game money) items within the game? On the one hand, this too provides practice of useful everyday math skills. On the other hand, if the item in question is a desirable one obtained by solving a particularly challenging math problem, then players who have difficulty with that problem could skip the math and still obtain the item. An obvious way to prevent this is to make such items "bind on acquisition," which means they cannot be traded or sold to other players.

Self representation. I discussed in Chapters 3 and 8 the value to mathematics learning that comes from the three identities: player, player-as-character, and character, particularly in terms of players' tolerance of mistakes. The Recife-effect you want to evoke in your game will be achieved by virtue of the player-as-character being in the game world and engaging in the various mathematical activities first-hand in the game world. A designer of a video game with an educational aim should be constantly aware of the three different identities and how they can be leveraged to achieve an educational goal.

For example, I suspect that a great deal of mathematics cannot be effectively learned within the game itself (i.e., by actions within the game world) if the game is intended to be first and foremost a game, albeit one with an educational mission, and not an "educational video game." For example, I do not think formal, symbolic-manipulative algebra can be learned in a game, though algebraic thinking certainly can. On the other hand, a designer could include some mathematical activities in the game that allow students who know algebra to apply it to their advantage, and in that way the game can motivate students to learn algebra and to reward them when they do. But learning symbolic algebra involves mastering a new level of abstraction, and that involves stepping back from the application for which the algebra is required, focusing on the new abstractions, and learning the formal rules to manipulate the symbols used to represent those abstractions. This inevitably takes the player out of the game action. Incidentally, this is as just true for the use of algebra to solve problems in the real world as it is in a game world.

What a game can do is provide players with lots of opportunities to handle concepts that lay the groundwork for algebra. In particular, the designer can ensure that the player is exposed to tasks that require recognizing numerical patterns and relationships, including functional relationships. A player who knows algebra can use it to complete such tasks, perhaps more quickly. But most players will solve the problem arithmetically rather than algebraically.

Where I see a real opportunity to leverage the game play to learn algebra is through "user generated content," where the player can design and build items to be used in the game. The MMO *Second Life* allows this, as does the social network game *Farmville*, but neither are games with stories in the sense of, say, *World of Warcraft*. A few genuine games have allowed limited use of user-generated content, but to date this has not been the norm. With current development tools it does present considerable challenges to implement, but we can expect to see this feature more and more in future games. The educational advantage it can offer is allowing players to carry out game-related mathematical tasks in their own time, free of the pressures of game action, yet within the overall game context. For example, if the game had an associated website linked to the game database, students could solve problems or build items on the website that will enhance their character in the game the next time they log on. This way, the powerful motivator of game advancement can spill out beyond the game to encourage players to carry out mathematical tasks in what I refer to as the "metagame" and what Gee calls the "big-G game." I'll say more about this in Chapter 11.

Tasks. Since most of the time that players typically spend in many video games is devoted to carrying out assigned tasks (e.g., quests given by an NPC in an MMO), task assignment is a primary means by which a designer can build curriculum into the game. It also allows the designer to provide integrated activities that require the player to combine different mathematical abilities and skills—preparing the way for genuine mathematical thinking and problem solving.

Teams. By presenting players with tasks that cannot be completed alone, a game designer can encourage the formation of teams of players. The teams can be enduring ones, as with the guilds that permeate current MMO play, or short-term ones put together on the fly to complete a single quest. Given the known educational benefits of collaborative learning, a good educational game should encourage teamwork. On the other hand, there are many people who greatly prefer to work alone, and some who find collaboration all but impossible, so the designer should make sure that a player who chooses to solo the game can do so, and still learn the same mathematics the other players do, though perhaps in a different order.

That last point about the order of the learning is worth a brief elaboration. With video game learning, different students will encounter curricular material in different orders. Assuming the teacher has structured the instruction to take advantage of video games, this means that the students will all be advancing in different orders and at different rates. This will require teachers to adopt new methods.

Time pressure. While I have to admit to a strong opposition to the use of timed tests as the sole means to determine students' mathematical abilities, such tests do have educational value in terms of encouraging and helping students to achieve the instant recognition and recall that are essential to further progress in the field. For example, a student who has not completely mastered the multiplication tables through to 10×10 will encounter difficulties in practically everything that follows in mathematics. The issue is not, as it is sometimes erroneously portrayed, that there is a need in and of itself to be able to recall the product of two single digit numbers. One rarely needs to do that in today's computer-rich society, even when shopping—you can always pull out a calculator. Where problems arise is when you have to resort to calculator use in the middle of a more complicated calculation, particularly if the issue is not multiplication of single-digit numbers but factoring a two-digit product. Such interruption of a difficult calculation to handle a sub-problem that should be a matter of instant recognition can cause you to lose your train of thought and fail to complete the calculation. A lack of total mastery of the multiplication tables can really drag a student down in more advanced parts of school mathematics and later at college.

Besides, in a video game, time pressure is not a negative feature but a source of considerable excitement that makes the game fun to play.

Worlds. The value of learning mathematics in a real-world environment was made clear by the Recife study. As I indicated in Chapter 3, virtual worlds provide an opportunity to take advantage of the "Recife factor" in the classroom or in the student's home. The degree to which the math skills learned in a real-world environment or a virtual world transfer to a paper-and-pencil classroom setting is another matter, which I'll address in Chapter 11. But many years of training professionals in simulators have demonstrated that skills learned and practiced in a simulated world may be readily applied in a sufficiently similar environment in the real world.

Getting the Math In

As Gee and others have pointed out, successful video games often involve activities that develop some mathematical skills; in particular, logical thinking, spatial reasoning, reasoning by cases, and general problem-solving ability. These are a fundamental part of game activity. But what about the meat-and-potatoes of middle school math: basic numerical skills and elementary geometry?

In my experience, doing this well (and that means without destroying the flow of the game) requires considerable thought, but the basic approach is clear. You look at the various constituents of the game and see how you can build in some mathematics. Those constituents are the world itself, the various artifacts in the world, the various characters in the world (players' avatars and NPCs), and the different actions that those characters can perform.

In *World of Warcraft* (and other MMOs), for example, players select a number of professions that their characters may pursue. In each profession, there are acquired skill levels, and at each level the character can acquire or manufacture certain kinds of items that may be used or sold to other players. Typical professions are alchemy, tailoring, herbalism, mining, blacksmithing, and armor crafting. Activities in such professions can naturally involve reasoning with whole numbers, fractions, multiple proportions, and (in cases such as tailoring or blacksmithing) elementary geometry and trigonometry. Thus, professions offer a wealth of opportunities to embed mathematics into the game. For the designer working on such a game, therefore, professions provide an abundance of low hanging fruit that is virtually irresistible.

It is, however, important to remember that for most players, the goal within a profession is to acquire or manufacture the item in question. Any math that arises is merely a means to that end. In fact, the designer should assume that the player has no intrinsic interest whatsoever in solving math problems. Thus, the math should arise naturally and should not present a major hurdle that causes the player to give up. It should not stand out as something that does not fit, a mere hurdle to be overcome; rather, it should seem perfectly natural to the player that he or she has to perform a certain calculation in order to complete the task at hand. In other words, the sensation for the player should be carrying out the profession task, not solving a math problem.

Pace is an issue here too, but it is the nature of the overall structure of MMOs that the pace in pursuing activities within a profession is much slower than in the more action-oriented parts of the game. The player accepts the slower pace because it is part of a larger activity that yields excitement elsewhere. For instance, when a warrior with blacksmithing skills crafts a new sword needed to slay a particularly dangerous dragon, the blacksmithing part is perceived as part of the build-up to the battle that will follow. As a result, the anticipation of the combat provides sufficient motivation for the much slower pace of calculating the various dimensions of the weapon and completing the manufacturing process.

One thing to be avoided is presenting the player with a task that he or she cannot complete. To achieve this, any mathematics that is required at the early levels of a profession should be extremely simple, the goal being to make sure the player knows what is required and can perform the required collection or manufacturing tasks. Then, as the player progresses in the profession, the math can be made increasingly more difficult, though only in small steps. Progress can be closely monitored by the game AI, which can provide plenty of opportunity to backtrack if the math starts to prove a significant obstacle.

The designer should avoid having the player end up sitting in front of the screen for a long period of time, stumped by—or struggling with—the mathematics. If that happens, then what began as a professional activity that the player wanted to do, and chose to do, and which made sense within the game, has suddenly turned into a math class, with all the accompanying attitudinal problems and psychological baggage.

Video games are about *action*. Most video games involve periods of thought and reflection, but only as an integral part of carrying out some activity with the player never losing the sense of performing that activity. If a player is manufacturing a cloak for her character, in the course of which she has to calculate the area of a piece of cloth, her answer to the question "What are you doing now?" should be "Manufacturing a cloak." If she replies, "Calculating an area," then you have a problem with the interaction design. This does not mean that it is not possible to put challenging mathematical problems into professions; rather, the trick is to do it the right way, with sufficient motivation and scaffolding for the player to succeed.

Of course, there will always be some players who by nature are happy to stop and reflect before acting. They are likely the ones who do fine under current mathematics pedagogy. For those students, the video game is simply an enhancement to their education that they don't really need, but won't do them any harm. In fact they will advance through the game faster than their classmates. I raced through the levels of *WoW* by being more reflective and analytic at each stage, developing good strategies.

The deeper into a game a player gets (the higher play level she or he achieves), the more this slow ramping can be relaxed. Advanced players can be, and from an educational perspective should be, presented with challenges that can be met only with considerable effort offline—effort that could involve looking up some math in a textbook or asking the math teacher. Remember, the aim is not to replace the teacher or the textbook; it is to supplement both, in terms of content and of motivation.

In many video games, quests constitute a main driving force that directs players' activities. Quests are highly motivational and players will generally expend considerable time and effort to complete them, with the reward being satisfaction (and something to brag about to fellow players) as well as the cool in-game items the quests can yield (money, armor, weapons, potions). The game designer makes use of quests to achieve a variety of results. For example, there are what are known informally in the game design business as "FedEx quests" that ask the player to deliver a message or a package to a certain NPC. The purpose of a FedEx quest is to get the player to discover new territory in the game world. They generally do not expose the player to any significant danger—unless she or he is sidetracked along the way. Another common form of quest is to collect a certain number of items that we'll call Q. These may involve fighting and killing monsters (for example if the Qs are wild boar tusks), but even if the Qs are items found lying on the ground, such as pieces of scrap metal, there is usually some element of danger. Quests of this kind provide the player both with practice at combat and a means to acquire experience points, money, and other possessions. A player who carries out such a quest repeatedly in order to acquire any of these forms of reward is said to be "grinding."

Most difficult—and dangerous—of all the quests are those that require the player to enter a dungeon in order to retrieve a particular object or rescue an imprisoned soul, for example. Dungeon quests often require the player to team up with one or more others in order to complete the task, which involves killing enemies—or groups of enemies—with powers far superior to those possessed by the player on his or her own.

A beginning game designer who wants to use quests to provide mathematics learning needs to understand the different kinds of quests in order to pick the most appropriate type in which to embed a particular learning experience. First, there are quests that are primarily about investigation—say of a building, a wrecked ship, or a region. As the action proceeds, the player may be provided with clues to assist the investigation. Spotting and acting on those clues can require some ingenuity on the part of the player. Additional clues can be added in real time if the game AI detects that the player has failed to make sufficient progress and is about to give up. If the investigation is of a geographic region, as in a typical FedEx quest, there is opportunity for tasks involving skills like spatial reasoning and route finding, speed and distance calculation, and determination of slopes.

One common device is to present the player with an obstacle such as a locked door or a locked chest that contains a crucial clue. In many games, locks are picked by a single mouse click, requiring only that the player has achieved sufficient seniority in the game to be able to pick such a lock, but there is no reason why opening the lock could not require some mathematics. With a little ingenuity, you can get a lot of middle-school mathematics into your game through the artifact of picking combination locks. Another obvious kind of quest that can require good mathematical thinking involves tasks requiring route finding, where the harder cases require finding the best route under different conditions. Such tasks are a feature of many games, but could be made much more challenging in order to develop mathematical thinking.

Investigation quests are frequently accompanied by a sense of imminent danger, since the investigation often takes place in a location where evil creatures roam. If the quest is to involve mathematics, the pressure of having to complete an action before a particular mob returns can be used to reinforce rapid recall skills, such as the multiplication table. Alternatively, if the mathematics requires some reflection, the player will have to be provided sufficient time to act, such as being provided with an invisibility cloak that shields the player from detection by an enemy until the problem is solved. The invisibility trick—variants of which *World of Warcraft* uses—has the important benefit that the player, while working to solve a math problem, continues to be part of the ever-changing game world. While older folks like me (and you?) might find multifaceted game activity distracting, to young players who have grown up in an era of multimedia multitasking, this is not a problem at all, and indeed its absence may be. The game designer has to remember that the player seeks action. In-game reflection is acceptable only when the player perceives it as being part of some ongoing action within the game.

In fact, the only limitations on embedding mathematical tasks within quests are (1) the task should seem natural within the particular game context, and (2) the player should find the mathematics doable under the prevailing circumstances. With regard to the second factor, it is fine if the player fails the first time he or she attempts a particular quest, or even two or three times, perhaps eventually giving up and coming back to try again later when he or she has gained more experience or acquired better equipment. The key is to make sure the player knows that eventual success is more or less guaranteed. In this regard, video game designers seem to do a much better job of securing and maintaining the trust of their players than do many school mathematics teachers. The success of video games in education depends upon making sure that implicit bond of trust between player and designer is not broken.

I could go on, but I suspect you get the general idea. The strategy is to look at existing video games, note the different features of the world, the player-characters in it, and the activities those characters typically perform, and ask what kinds of mathematical tasks the player-character can be presented with during the course of game play. In doing this, the designer has various ingredients to play with.

I'll give just one specific example. It's an obvious one, but it should serve as a good illustration. Video games are littered with treasure chests the player has to open to acquire their contents, often by solving a puzzle. Obviously, in a math ed video game the puzzle should (sometimes) involve mathematics. For the sake of argument, suppose a region of the game world were full of chests that required the player to factor numbers to open them. The player keeps encountering such chests, getting better and better rewards, but on each successive occasion has less time to open the chest before the evil monster returns and she finds herself in a fight for her life, thereby forcing her to improve her ability to do the math more efficiently. Surely, it will not be long before your player is a factoring wizard. What's more, she will have achieved her mastery without anyone ever having uttered the baffling phrase "factor into primes," let alone the decidedly more off-putting instruction, "And now, class, we're going to learn about prime factors." Instead, she will have *discovered* it. In a natural, "real world" setting. On her own.

(Well, not exactly on her own. The game designer would have crafted the lock-picking interaction to give her the *illusion* that she was engaged in a process of free discovery.)

What's more, she will have become a factoring wiz at her volition.

(Well, again not exactly. *Her* volition was to get into that really special chest and grab the Jewel of Eternity. The designer took that goal and forced her to jump through some mathematical hoops to get there. Not that it seemed to her like jumping through a hoop; she was just doing what arose naturally in the game at that point.)

Notice that the dreaded "M" word never comes up in this interaction, nor does the designer ask the player to grab a paper and pencil and "do the math." (Though she may do so—at her own volition, of course.) Sound good? Well, it could be made to work, but there are some problems to overcome.

One problem is that the chests are persistent features of the game world, shared by all players. Only one player—out of the thousands who could inhabit the world—can work on a particular chest at any one time. To be sure, in games like *World of Warcraft*, there are quests (for example, escort quests) that likewise can be tackled by only one player at a time. But they generally take only a few minutes to complete and they are almost entirely physical, which means that waiting players can either follow behind and observe how the quest unfolds, or even get involved in the action by helping the current player, usually in exchange for a return favor on the next run. In these situations, the player waiting to tackle the quest does not have a sense of halting the game play while someone else completes it. Indeed, it is interesting to watch another player try to complete the quest. But it could take some players a considerable length of time to factor the number correctly to open the chest, perhaps even ten minutes or more, and for much of that time the only real "action" is taking place is at the neuronal level inside that player's head. (Indeed, that's what, from an educational perspective, we *want* to happen!) There would be little for waiting players to see or do. No action to observe or participate in. No way to learn in advance how to solve the problem. Moreover, you would likely see the emergence of the educationally disastrous scenario of waiting players using the chat channel to harass, intimidate, and ridicule the player working on the chest in order to force them off of it. Before long, the player-created game websites that will spring up will be full of advice to stay away from what will inevitably be an interminable and boring wait, perhaps followed by a psychologically harrowing period of being "flamed" on the chat channel.

Clearly, that approach will not work, except in what is known as a dungeon instance. MMOs typically provide quests for individual players or small groups that are carried out in an environment isolated from all other players. The game environment is literally *instanced* for that one player or group. The designer could put some chests in game instances, but that loses some of the educational benefits of a multiplayer game.

With mathematical challenges embodied in persistent world artifacts out of the question, it is clear that any mathematical manipulatives will have to be in the form of devices individual to each player; for example, an all-purpose lock-picking device. Such a device could be obtained from a quest giver, or acquired from a nearby NPC, such as a drop from a mob or a purchase from a vendor, or found in a handily placed tool chest. There are other possibilities.

A number of educational games adopt this approach of the player's avatar using some device that involves some curricular activity (for example, *Dimen-sionM*, mentioned in Chapter 1), and they all have one thing in common: HUDs.

HUDs

It is the essence of many mathematical tasks that they take time. That is true in a video game every bit as much as in the real world. Eventually, the player has to stop acting and start thinking. Perhaps not the first time she attempts the task, when she uses trial and error. Occasionally she will strike it lucky. But it won't take long before she recognizes that there has to be a quicker way, one that is more efficient over the long haul when many more similar challenges are encountered. And of course, there is! The player might think of it as acting smart; I call it doing math.

Once the player starts to use mathematical thinking in trying to complete a task, her attention is entirely on the particular puzzle that is the key to the task. At those moments, the physical and story-advancement aspects of the game come to a temporary halt. The puzzle—more precisely the device used to solve the puzzle—can thus be built as a 2D screen overlay, known as a *head-up display* (HUD). See the example in Figure 5.

Since a HUD is two dimensional, it is quick and easy (and cheap) to build. In particular, this means it is easy to make many versions of the same puzzle with the same device, each one varying in some educationally important way from the others. It is also a simple matter to tweak a particular puzzle if you find that it is not working for the players in the intended way. Yet another benefit is that it becomes

Figure 5. A HUD in the game *Escape from Frankenstein's Castle*, by Spark Plug Games.

a relatively easy task to craft versions of the game suitable for players of different mathematical ability levels.

What I think is crucial in using HUDs in this way is that the mathematical tasks carried out using these devices should never look and feel as if they are outside the game action or have been added on to the game. They must be—and must be perceived by the player to be—a natural, believable, and integral part of the game action. This is one reason why it would be difficult to take an existing game, perhaps one whose commercial shelf life has passed, and build in the mathematics, as some people have proposed. A key feature of an effective mathematics education game is that the back story and the game world are both developed along with the in-game curriculum. If you give up the close integration of the mathematics into the game, your students will lose all of the advantages of learning everyday mathematics in a natural environment like the Brazilian street market.

These considerations mean that not only must the HUDs have a look and feel just like the rest of the game world, their existence and use have to be embedded in the game environment. When a player's avatar pulls out a lock-picking device, for example, the door to be opened should be visible "behind" (more precisely, around) the device. There should be no sense of leaving the game, even momentarily. In particular, recall from the research of Jean Lave and Deanna Kuhn that I discussed in Chapter 2, when supermarket shoppers are presented with artificial shopping problems just outside the supermarket, their performance level drops considerably. Doing the math has to feel like—indeed, has to *be*—just another part of the game. It should not even feel like "doing math" in the familiar sense, though even I have to acknowledge that this may be too much to hope for. Perhaps the most you can expect is that in your game environment, although the players know they are being asked to solve a math problem, they view it and approach it differently from the way they do in math class. The Recife study is a good example of this.

One huge benefit of embedding mathematical tasks in HUD devices is the natural facilitation of repetition. Let me elaborate, and with as much force as I can muster.

How Do I Get to Carnegie Hall?

You've surely heard the joke. The elderly lady in New York goes up to the policeman and asks, "Officer, how do I get to Carnegie Hall?" The policeman looks back with a smile and says, "Lady, there's only one way: practice, practice, practice." That is as true for mathematics as it is for being a concert pianist.

Contrary to what you may have read, much of mathematics is *not* a natural way of thinking. Some mathematicians claim that they have always found it natural, and maybe some have, but there is no reliable way to test the veracity of such a claim. The most that we can conclude is that some people, once they have mastered mathematics, do find it natural and cannot recall ever finding it otherwise. But then, we all find walking natural and cannot remember ever finding it otherwise, yet every parent knows how much effort it takes a young child to master the skill. There is also a significant difference with mathematics. We do not have to be taught to walk. We pick it up, driven by an instinct. But we have to be taught mathematics, and it takes time and effort to learn it. I'm skeptical of people who say they *always* found math easy or natural. Quite frankly, while I find mathematical thinking an entirely natural activity, I also find it impossible to reconcile the time and effort required to master new mathematics with claims that it is intrinsically natural. As it happens, I do recall a time when I did not find mathematical thinking natural. I was the last child in my elementary school class to master my multiplication tables, and I struggled with the subject until about nine or ten years of age. I was well into my teens before it began to seem natural.

Innately natural or not, the only way we can master mathematics—and almost any other activity for that matter—is through repetitive practice. There is no silver bullet or magic pill. This is as true for mathematics as it is for skateboarding or playing the piano. As Malcolm Gladwell pointed out in his book *Blink*,[2] experts in any domain have typically spent at least 10,000 hours attaining that level of performance, regardless of the activity. (Interestingly, this is the total time over 95% of today's youth will have spent playing video games by the time they graduate from high school.) Of course, expert-level mathematics performance is not the goal for an educationally oriented video game. But attaining any target level of performance requires practice. The challenge facing the educator is to design the repetitive practice so that at the very least the learner will not give up, better still the learner will remain self-motivated and ideally will enjoy the process. Most of the critics who decry repetitive practice in mathematics education are really arguing against its implementation in an unmotivated, mind-numbing way that puts many learners off, often for life.

This is where video games can exert real muscle. As every parent knows, one of the most dramatic features of video games is the amount of time players typically spend on repetitive tasks—tasks that can seem unbelievably boring, and which players will sometimes say *are* boring. Indeed, the capacity to drive repetitive practice is the main pedagogic principle behind virtually every math ed video game on the market today. Those games entice players to spend hours performing repetitive tasks in order to carry out fairly traditional "drill-to-skill" mathematical exercises, not unlike the many Saturday morning "math skills" classes offered by the private education sector.

---

[2] Malcolm Gladwell, *Blink: The Power of Thinking without Thinking*, Little, Brown and Company, Boston, 2005.

Though I have never argued against the benefit of repetitive practice, it would be a dreadful waste of the enormous power of video games to use that power solely to develop basic skills mastery through "time-on-task" repetitive practice. Practice can make people better at higher-level mathematical thinking as well. It's just harder for the video game designer to incorporate appropriate game play to achieve it.

For example, repetition of tasks that develop genuine problem solving ability would be extremely difficult to achieve with a persistent, shared device such as a locked chest. In order for a player to gain mastery of the various associated mathematical skills required to open the chest, the player would have to keep encountering the chest, each time in a different numerical starting configuration, and perhaps similar chests of greater difficulty in different game world locations. This would clearly present the game designer with a difficult task of making it natural—and exciting—for the player to keep opening the same kind of chest, or even the very same chest in the same location, and ensuring that he or she had a natural motivation to do so in terms of progress through the game.

It's hard to imagine even the most creative and imaginative game designer and back story author creating scenarios that would result in a player carrying out five or six repetitions at one stage in the game, in one location. On top of which the designer would have to ensure the player keeps encountering the device or virtually identical devices at later stages of the game in order to provide the curriculum revision and topic-integration activities necessary to achieve good learning. But with mathematics tasks embedded in devices the player either wears, carries around, or can acquire on-site when required, it is possible to require the player to use the same or similar devices many times, at different locations, under different circumstances, and with increasing levels of mathematical difficulty to achieve different specific game objectives.

In short, based on everything we know about player behavior in video games, it seems clear that embedding mathematical tasks in player-held devices can drive regular and repeated practice both of basic mathematical skills and more substantial problem solving ability. Indeed, such tasks might even be less tedious than killing enough boars to collect twenty tusks when only one in five boars turns out to have a retrievable tusk—a fairly common sort of requirement in *World of Warcraft*.

As a testament to the fact that video game players will voluntarily do the math if it makes a difference in the game, the Wiki-style websites that spring up around the major MMOs are full of player-generated statistics about the different "drop rates" at various locations. In many cases, such comparison tables are accompanied by one player explaining to another why a mean (average) drop rate does not guarantee such a rate on a particular occasion. What was that newspaper headline about video games preventing students from learning?

There is an additional benefit to embedding the math in artifacts under an individual player's temporary or permanent possession as opposed to in a persistent object in the shared game world, such as a locked chest. Let me explain. Although it is possible to make use of persistent, world-based devices in instanced dungeons, those locations are generally not designed to be played by a single player. Their purpose is to encourage—actually force—teamwork. Accordingly, they are built so that a single player cannot survive for more than a few moments in such an environment. Instead, they are restricted to a small number of players, who must form a team, or party, in order to "run" the dungeon. In order to succeed, players must assemble a party with an appropriate balance of different skills and strengths, and plan and coordinate their actions. In the case of a mathematical challenge such as a locked chest, if it is encountered in an instanced dungeon, it will inevitably be that one person in the party does the math and solves the problem. That player will surely get considerable educational benefit from the activity. But the others will get little or nothing; they will be mere spectators.

With player-owned objects, however, this will not arise. Even if the players are working together as a group, each individual player must operate his or her own unit. The data presented by the unit can be randomized for each player, so it won't work for one player to solve the problem and simply tell everyone else what to do next. Of course, one player could explain to the others the general *method* required to solve the problem—but that is a highly desirable outcome.

What about the Game? Engineering Play

So far I have been talking about building the world, both the environment and the artifacts put in it. But the game is something else. The game is what emerges when people enter the world and start to play. How does the designer engineer the actual play? I should stress that the design and construction of the world and the objects and characters that populate it should not be carried out separately from the design of the game. The two processes need to be tightly interwoven, since decisions in each can have major impact on the other. But for the purposes of describing the process, it is convenient to carve it up in this way.

In designing the game, the designer has to ask several interrelated questions:

1. In the case of an educational game, will the intended player be in a classroom, under direction of a teacher, or playing at home for recreational purposes?
2. Assuming the game is not designed for exclusive classroom use, why does a particular player decide to play the game? What are his or her personal goals?

3. Does the player prefer to play alone, with one companion, or in a group?
4. What kinds of activities in the game does the player prefer? What does he or she really not like to do and will avoid if at all possible?
5. How and when does the player use the math-embedded artifacts?

I'll start with question 1. For a game intended as a classroom resource, the game elements built into the system can be fairly minimal, though they do not have to be. Perhaps what a teacher really wants is not so much a game as a classroom resource, something akin to a textbook or a school laboratory. The teacher will design and organize the learning activities the students will carry out in the "game" environment. Those activities may be structured as a game, but it will be the teacher's game, not the game designer's. There is enormous potential in such uses of immersive environments, whether used in single-user, small group collaborative, player-versus-player competitive, or MMO mode.

Turning to question 2, why does a particular player decide to play the game and what are his or her personal goals? I said something about the reasons people play and what they look for in a video game at the start of this chapter. Let me make some additional remarks.

There have been a number of studies of what motivates players to play, and continue to play, a video game, though the game studies field is so young that our understanding is continually changing. In 1990, when online role-playing games were predominantly text-based, game researcher Richard Bartle identified four different kinds of online fantasy game player (the names I use for Bartle's four categories are my own):

- *problem solvers*—players who seek to succeed in specific tasks within the game;
- *explorers*—players who gain the most enjoyment from discovering as much as possible about the game world;
- *socializers*—players whose primary interest is socializing with other players; and
- *competitors*—players whose main goal is to compete with and attain power over other players.

Bartle's taxonomy, while imprecise, is well known within the game developer community. The intention is not for the game designer to build in elements that explicitly and exclusively appeal to each kind of player separately, rather that it highlights the fact that players have very different reasons for playing and the game design should allow for this diversity. In particular, the taxonomy evidently provides a general perspective that helps us address not just my game design question 2, but questions 3, 4, and 5 as well.

I'll have more to say on questions 2, 3, and 4 as we continue. My main focus now, however, is question 5. The first thing to note is that a player will use a math-embedded artifact because its use is required in order to advance in the game. Well, that seems rather obvious. But as I have observed earlier, players have different motivations to play a game, and they can approach it in different ways that can affect the way they react to a math task.

In particular, recasting the Bartle taxonomy, a multiplayer game with a virtual world, such as *World of Warcraft,* facilitates three different kinds of play action (in this context I am not counting socializing with other players as a play action):

- exploration (of the world, the creatures in it, or their history or purpose),
- competition (against other players or NPCs), and
- problem solving (within the world).

Math-embedded artifacts and other forms of mathematical challenges can be geared toward each kind of action. But what should they look like?

Problem solvers. Though I listed problem solving last above, I'll deal with it first because studies of gamers show that problem solving per se motivates relatively few players. That is not to say that they do not enjoy solving puzzles. Indeed, for the majority of players, puzzle solving constitutes an important part of what makes the game fun to play. Take away the puzzles and the game would lose much of its appeal, perhaps too much, leading players to get bored and abandon it. But almost all players approach the puzzles as obstacles to be overcome in pursuit of a greater goal in the game. After all, if all a person wanted to do was solve puzzles, why waste time playing a complex video game? Why not go out and buy a puzzle book, or a computer-based problem set?

Often, the puzzles are along the lines of police detective work, looking for and piecing together clues, either in the local area (of the game environment) or in the quest text. There is no reason—at least no insurmountable reason—why those puzzles cannot involve mathematics. The important thing is that the challenge should arise naturally and make sense within the game.

Actually, arising naturally is just one of two important things about game puzzles. As I observed earlier, one of the things that good games do well is develop in players a belief that, no matter how difficult a puzzle might seem at first, everything they need to solve it is provided somewhere, and with clues as to where. Thus, players become confident that if they persist long enough, they will eventually arrive at the solution. In presenting players with a mathematical challenge, it is important to follow this critical design dictum, and not break the player's trust that he or she has not been asked to do something beyond them. That is, unless they are attempting to operate at too high a level in the game for their current game ability. Players are fully used to the fact that a task that is impossible when their character is at level 17 can become just doable at level 18 or 19, perhaps with a little help from another player, and can be handled with ease at level 20. One of the key design issues for putting math tasks into the game is to arrange them by levels—though players are likely to differ somewhat in what game level they will be at when a particular math task becomes doable.

Explorers. For a player whose motivation is primarily exploration, a mathematical challenge may be nothing more than a means to an end. The puzzle simply presents an obstacle to entering that next room or crossing that bridge to the adjacent geographic region. The fact that the player has to do some work to achieve that goal is what gives the sense of accomplishment when he or she succeeds, and is what makes advancement in the game something to brag about to others in the game. Thus the puzzle is an important constituent of the game play that cannot be left out. The player is likely to persist until the problem is solved, provided that

- the puzzle makes game sense in terms of achieving the player's game goal,
- the puzzle is presented in a way that keeps the ultimate end in sight,
- the puzzle is perceived as doable, and
- solving the puzzle does not take the player cognitively out of the game.

The last criterion here is the one that can present the game designer with the greatest challenge. Puzzles in many games typically require the player to continually interact with the environment, with NPCs or other players or both. Thus, the play action (in this case, exploration) continues. If the activity required to solve the puzzle comes across as separate from the play action, however, then even if the set-up makes the puzzle seem natural, a player who gets no intrinsic pleasure from problem solving may decide to seek advancement in the game by an alternative game path.

Competitors. You might think that a video game designed to appeal to competition seekers would not need puzzles at all, but in fact most such games have them. Presumably this is because players are not exclusively seeking competition. When you look more closely, even fast-action, first-person shooters usually have puzzles embedded in them—figuring out how to use this new weapon, locating the ammunition, deciding where to commence the attack, choosing a route that avoids detection by the enemy. These tasks are presumably required to give variation in game play. Even the most ardent "shoot 'em up" player would rapidly get bored with an endless fight sequence. The puzzle solving episodes provide an opportunity for tension to build in the player.

Many games, including all MMOs, provide far more than just competition, of course, and we can assume that a player who has chosen to play such a game has done so for a good reason. Experienced MMO players who are primarily motivated by competition are well accustomed to acquiring powerful weapons, armor, and magic spells prior to going into combat, and this provides one avenue to embed mathematical tasks. Provided the payoff from solving a mathematical task is sufficiently great, they will persist—though again, it is important that the challenge makes sense in terms of the reward it yields. Asking a player to solve a quadratic equation in order to obtain a Sword of Doom is unlikely to be successful unless the game designer can find a way to make that equation highly relevant to that weapon—something that I have great difficulty imagining, though admittedly I am not an experienced game designer.

Regardless of the type of play action that the player prefers, the overriding goal for the designer of a math-embedded game who wants that game to be successful *as a game* is to make sure the math puzzles (indeed, all the puzzles) do not break the rhythm of the game, and that they enhance and enrich the game experience.

While on the topic of not breaking the rhythm of the game, for a game with an educational purpose, the designer has to make sure that the desire to succeed in the game—which we can assume is the player's primary motivator—is not crushed by an inability to complete a particular task (which is the math ed game designer's goal). This is where Gee's Principle 16[1] comes into play, implemented into the game design as I indicated in Chapter 8. Again, in order to ensure player-flow within the game, the designer should ensure that at any moment of play, the player's mindset is that of his or her avatar, striving to achieve a goal within the game, not that of solving a math problem exterior to the game. When asked "What are you doing now?" the player's answer should be in terms of some game goal (rescue the princess, slay the dragon, etc.), not trying to solve some math problem. It may be impossible to achieve this in all cases with all players, but it is a goal that should guide the game design.

[1] The Multiple Routes Principle: There are multiple ways to make progress or move ahead. This allows learners to make choices, rely on their own strengths and styles of learning and problem solving, while also exploring alternative styles.

6+5 Algebra and Beyond

I hope my enthusiasm for the potential of video games in mathematics education has not led any reader to assume they are the answer to all our current math ed woes. They are not. For one thing, although they do provide an ideal learning environment for much of what I have been calling everyday math—an environment that many young people find attractive—they are not for everyone. Some people simply are not interested in video games and don't want to play them. (About 3–5% of the target age-range audience, according to the latest surveys.) I am not suggesting that an educational video game is intended to replace all, or indeed any of, other forms of mathematics education. It is meant to be an optional addition to everything else that is available, including school. To be sure, the option is not necessarily that of the student or the parent. It can be exercised by the teacher or the school district. If either chooses to use a particular video game, the student will have no choice, even if they never play video games at home. They don't have a choice about taking the math class either.

I should also stress again that while video games are ideal to help students learn *everyday* math, they are not so well suited to provide learning for other kinds of mathematics. In particular, because their underlying educational philosophy is situated learning, they do not lend themselves naturally to teaching abstract, symbolic mathematics, such as algebra. At least not in the way I have been describing. With a different approach, however, I see no reason why you could not design a video game to help students learn algebra or any other topic from K-12 mathematics. And if you expand your horizons from the game itself to the entire "metagame" (Gee's "big-G game"), then I see no limit to the mathematics that could be learned with the aid of a suitably designed video game. The issue is what role the game plays in the educational process.

153

Later in the chapter I'll describe some ideas to use video games to address algebra and other topics. Before I do that, however, I want to set video games aside for a moment and take a broad look at some of the issues that you encounter as soon as you try to go beyond *situated* learning of *everyday* math (i.e., learning and use of everyday math in a real-world or world-like environment) to learning abstract, symbolic math *in any manner*, whether from a textbook, at home, in the classroom, with a private tutor, in a video game, or otherwise.

Two Kinds of Math

As the Recife study and others like it have demonstrated so dramatically, abstract, symbolic math—the kind usually performed using pencil and paper—is a very different mental activity to what I am calling everyday math performed mentally in a real-world or simulated world-like context. To think of the former as simply a representation on paper of the latter is misleading and wrong. It's an understandable mistake to make—as I did for many years—for those of us who mastered symbolic math at an early age. For once you have such mastery, symbolic math does indeed seem to be a representation on paper of basic math performed mentally. Perhaps by that stage it is for some people. But research suggests that for most people symbolic math is not a representation of everyday math, at least in the beginning.

I should emphasize that the point I am making here is a practical one about what is involved at a conscious level in learning and doing mathematics. How the brain does mathematics is another thing. In their book *Where Mathematics Comes From*,[1] the cognitive scientists George Lakoff and Raphael Núñez present a theory of how the brain develops mathematical capacity step-by-step from early childhood experiences in the world and ending with understanding the famous Euler identity $e^{i\pi} = -1$.

Lakoff and Núñez do not claim that each step of the process, the creation of new "metaphors" in their terminology, is conscious or deliberate. Their main goal is to understand how the circuitry of the brain can adapt to produce new mental capacities. Thus, what they suggest is not at odds with the point I am making about there being two distinct kinds of mathematics. As it happens, however, I along with a number of other mathematicians, find that while I am in broad agreement with much of what Lakoff and Núñez say in the early part of their book, where they cover basic mathematics, I part company with them when they get to abstract, symbolic mathematics.

My point of departure from Lakoff and Núñez is that many of the concepts of advanced mathematics (roughly, calculus and beyond, though the transition oc-

[1] George Lakoff and Rafael Núñez, *Where Mathematics Comes From: How the Embodied Mind Brings Mathematics Into Being*, Basic Books, New York, 2000.

curs prior to that) are linguistically constructed, and have no natural real world meanings. This is not to say that the concepts cannot be applied to the real world. Indeed, in many cases that is precisely why they were developed in the first place. But those and other advanced math concepts are created through the symbols used to represent them. In my view, you cannot, as Lakoff and Núñez claim, effectively construct them from more basic concepts. Their meanings have to be bootstrapped within mathematics, and that means there is no alternative to mastering them than to first learn the formal definitions and the symbolic manipulation rules, then use them repeatedly—at first without understanding them—in different mathematical contexts, until their meaning emerges.

Differential calculus provides an excellent example. The derivative of a function, for instance, has no natural meaning. True, high-school teachers often try to circumvent this difficulty by telling students that the derivative is the slope of a graph. But that is just plain false, and I do wish they would stop saying it, as it leads to difficulties later on. Yes, you can *use* the derivative in order *to calculate* the slope, but that is just one way to use the derivative, it's not what the derivative is.

You might have noticed my use of the modifier "effectively" when I said that "you cannot, as Lakoff and Núñez claim, effectively construct them [the objects of abstract, symbolic mathematics] from more basic concepts." With proper guidance, it may be theoretically possible to progress through all of mathematics in a fashion that allows understanding of each new concept or method before moving on to the next one. Thus, in a strict sense, though not a practical one, Lakoff and Núñez may be correct in their basic assumption. In viewing mathematics as subdivided into two distinct subclasses, everyday math and symbolic math, I am making a distinction that is more one of practice and practicality than intrinsic to the subject. The goal of this book is to provide mathematics educators with practical advice on mathematics education, not to add to the literature on the philosophy of mathematics.

The problem with basing a pedagogy on Lakoff and Núñez's theory is one of time. Even if it were possible in principle, it would take far too long. For instance, the university student of physics, engineering, or economics needs to be able to solve differential equations by the end of the freshman year or they will not be able to make any progress in their main subject. They do not have the time to acquire conceptual understanding of what they are doing. The most that is possible in the limited time available is to achieve some degree of what I have called *functional understanding*; namely, knowing when and how to apply each technique and what its limitations are.

The "learn-by-the-rules approach" works because the human brain is a naturally evolved symbol processor. We are linguistic creatures and can learn symbolic processing rules, without any need (other than perhaps a psychological desire) to understand what they mean. Hence generations of students have learned how to compute derivatives of functions, and to make effective use of that ability in their subsequent careers as scientists, engineers, or others. They frequently do so without ever understanding what the derivative means. To them, calculus is simply a tool. They know when and how to use it and that provides sufficient meaning for successful use, but from a mathematical viewpoint the real meaning remains hidden.

In fact, as I noted in Chapter 9 (page 108), even in the early stages such as learning numbers, the symbolic representation precedes acquisition of the concept. But in such cases, the concept comes directly from the familiar real world by a single-step abstraction, so the learning process is different. An analogy would be that learning to count collections and acquire the number concept is like learning to swim by stepping into the shallow end of the pool and moving slowly into deeper water, at first taking one foot off the ground, then taking small swim-like hops until you no longer need the support of the ground. In contrast, learning calculus is like being thrown in the deep end. There is no support from the ground. You either learn to swim by performing the actions the instructor showed you, or else you sink. Unfortunately, for many students this metaphor is all too apt for calculus instruction.

Thus, calculus is in many ways a *cognitive technology*—a tool you use without knowing much, if anything, about how it works. For example, few people know how an automobile engine or a computer works, but that does not prevent those people from becoming skillful drivers or computer users. Successful use of a technology generally does not require an understanding of how or why it works.

It is because you can calculate derivatives, integrals, and the like simply by following symbolic rules that we refer to the method by the name "calculus." The word is Latin for "pebble," and its use in modern mathematics reflects the fact that in Roman times calculation was performed by manipulating pebbles in the sand according to a set of prescribed rules. In a typical college calculus course, students learn to follow certain symbolic rules and when and how to apply them. What those rules mean is the topic of another course, generally called Real Analysis, a subject usually taken only by mathematics majors, and only after they have demonstrated mastery in calculus.

The distinction I am drawing between everyday math that is directly abstracted from the world and abstract symbolic math that is linguistically created is complicated by the fact that even the most elementary parts of mathematics can be done in an abstract, symbolic fashion. For that reason, they may also be taught that way—and for many generations that is largely how they have been taught, and in many parts of the world still are. Because of the way our brains work, given enough repetition almost every one of us can master basic skill at symbolic math to the point where we can pass a test. But that does not guarantee understanding, nor does it imply the ability to take that skill and apply it in the real world, or indeed in another mathematical context. This is why today's more enlightened mathematics educators insist, correctly, that students achieve more than the ability to mimic a ten-dollar calculator—but that they also acquire conceptual (or at least functional) understanding and the various other strands of the NRCR's notion of mathematical proficiency.

Even if you are not convinced by the above points, however, and maintain, along with Lakoff and Núñez, that mathematical activity is a continuous spectrum that with the correct instruction could be traversed in reasonable time, then given that such instruction has never been achieved, the pragmatic approach for the educator (and the educationally-motivated game designer) is to avoid being sidetracked as to the exact relationship between everyday math and symbolic mathematics, and simply view them as two different kinds of mathematics. Which is what I am doing in this book.

What about Symbolic Math?

I have tried to convince you that video games provide an ideal medium for learning how to do everyday math. But what about symbolic math? Can that be learned with a video game? If you are a parent, you probably have a particular interest in this question. Yes, it's clearly valuable for everyone in modern society to have a good mastery of everyday math—the mental stuff, carried out in a real-world environment. But doing well in school and passing exams is important too. And that involves being able to do abstract, symbolic math, in particular high-school algebra.

To be sure, there are some commercial games on the market already that claim to teach algebra (e.g., Tabula Digita's *DimensionM*), as well as a number of attempts in universities to develop non-commercial games that do the same. But to my critical eye, while all can be said to have some merit—and are surely not a waste of time—none of them has come close to achieving the full educational potential of the medium. One problem is that they come across as classroom math grafted onto a video game. Students might prefer playing such a game to having a regular math class (they often say that), but they won't choose to play such a game on its merits, and faced with a choice between such a game and, say, *World of Warcraft*, they will opt for the latter. Moreover, the learning those games provide is very shallow and restricted to recollection of basic facts and some practice of skills rather than mathematical thinking.

Still, lack of success so far does not mean that no one will find the right recipe some time in the future. But I suspect not. I think it is unlikely anyone could develop a video game that has symbolic mathematical activities embedded in it (in particular, algebra) and which works well as a game. Using video games to help students learn symbolic math requires a different approach that I'll turn to momentarily. Before I do so, however, I need to explain what I—and my fellow professional mathematicians—mean by algebra. The reason I need to do this is that experience has taught me that many people do not really know what it is, including, I regret to say, many math teachers and not a few college professors.

Algebra: What Is It Anyway?

Algebra is not doing arithmetic with one or more letters that denote numbers, known or unknown. For example, putting numerical values for a, b, and c in the familiar formula

$$x = \frac{-b \pm \sqrt{b^2 - 4ac}}{2a}$$

in order to find the numerical solutions to the quadratic equation

$$ax^2 + bx + c = 0$$

is not algebra, it is arithmetic.

   In contrast, deriving that formula in the first place is algebra. So too is solving a quadratic equation not by the formula but by the standard method of "completing the square." Here, in a nutshell, is the distinction. In arithmetic, you carry out numerical, *quantitative* reasoning *with* (specific) numbers, and the answer is a number. In algebra, you engage in analytic, *qualitative* reasoning *about* numbers (in general), and the answer is generally an algebraic expression (though it can be a number). The key distinction is the kind of reasoning, not the kind of answer. In arithmetic you proceed from the known to the unknown. In algebra you use the unknown to proceed to the unknown—you acknowledge your ignorance by denoting what you do not know with a symbol, then you express what you do know.

   What causes many people to be confused between the two is that when we are doing arithmetic we sometimes use letters to denote specific numbers (as when we apply the quadratic formula), while it is possible (though these days not common) to do algebra without using letters at all. The ancient Greeks and the European mathematicians of medieval times did algebra using not letters but words—what is nowadays sometimes referred to as "rhetorical algebra." What distinguishes algebra from arithmetic is the kind of reasoning used. In particular, algebra involves a particular kind of logical thinking (e.g., about relations, inverses, and equivalents between and within quantities). It is thinking logically rather than numerically. Ultimately, algebra, at least at the high-school level, is still about numbers (strictly speaking, generalities about numbers), but that does not make it arithmetic.

I believe that students' failure to recognize this distinction (in many cases because no one points it out to them) lies at the heart of why many of them have so much difficulty mastering algebra. Many students, when they first encounter algebra, try to solve problems by arithmetical thinking. That's a natural thing to do, given all the effort they have put into mastering arithmetical thinking. At first, when the algebra problems they meet are particularly simple (the teacher's classification as "simple"), this approach works. In fact, the stronger a student is at arithmetic, the further they can progress in algebra using arithmetical thinking. For instance, many students can solve the quadratic equation $x^2 = 2x + 15$ by trial-and-error arithmetic, without using any algebraic thinking at all. The possibility of reliance on arithmetic skill leads to the seemingly paradoxical situation that students who are better at arithmetic may initially find it harder to learn algebra. To do algebra, for all but the most basic examples, you have to *stop* thinking arithmetically and learn to think algebraically.

A good example of algebraic thinking is when you want to write a macro to calculate the cells in a spreadsheet like Microsoft Excel. With a spreadsheet, you don't need to do the arithmetic; the computer does it, generally much faster and with greater accuracy than any human can. What you, the person, have to do is create that spreadsheet in the first place. The computer can't do that for you. It doesn't matter whether the spreadsheet is for calculating scores in a sporting competition, keeping track of your finances, running a business, or figuring out the best way to equip your character in a video game. You need to think *algebraically* to set it up to do what you want. You have to think *about* or *across* numbers *in general*, rather than in terms of specific numbers. This one example alone makes it clear why, in today's world, logical, analytic thinking across numbers—algebraic thinking—is so important. It is precisely because qualitative, analytic reasoning plays such a major role in today's society that former President Bush's National Mathematics Advisory Panel recommended in 2008 that all United States high-school students become proficient in algebra.

Incidentally, I think that spreadsheets can provide today's students with more meaningful and fulfilling applications than the problems about trains leaving stations or garden hoses filling swimming pools that my generation had to endure. To my mind, all school algebra textbooks should be full of spreadsheet exercises, but to date this is not the case. In fact, many of the examples and problems you find in beginning algebra textbooks are actually not algebra but "arithmetic with letters." The intention behind such problems is to help students make the transition from arithmetical thinking to algebraic thinking. Personally, I have never been convinced that such an approach works. It's like using training wheels to teach a young child to ride a bike. Riding on two wheels is fundamentally different from riding on four, and no amount of the latter can teach you how to do the former. The most that the training wheels can achieve is perhaps help the child gain a bit of confidence. Once those training wheels come off, however, the child still has to learn how to keep upright on two wheels. A more efficient and much more fun way to achieve that feat is the way I learned, with my father jogging alongside me with his hand on the back of my saddle to help me maintain balance and give me encouragement and confidence that I could do it. That is how I taught my children to ride bikes. Likewise, I think the best way to teach students how to think algebraically is for a human teacher to work with them, at the same time giving them encouragement and confidence in their abilities as they learn to think in that new and different way, not by using "arithmetic with letters" as training wheels. This approach would help the students to realize from the outset that learning algebra is a difficult cognitive task. The human brain does not find it a natural way to think; it requires training. The main problem is the level of abstraction. Numbers are abstractions from the familiar everyday world, but algebra focuses on abstractions across those abstractions. Let me elaborate.

Although many people find arithmetic hard to learn, most of us succeed—or at least pass the tests—provided we put in enough practice. What makes it possible to learn arithmetic is that the basic building blocks of the subject, numbers, arise naturally in the world around us—when we count things, measure things, buy things, make things, use the telephone, go to the bank, check the baseball scores, etc. Numbers may be abstract—you never saw, felt, heard, or smelled the number 3—but they are tied closely to all the concrete things in the world we live in. With algebra, however, you are one more step removed from the everyday world. Those x's and y's that you have to learn to deal with in algebra usually denote numbers in general, not particular numbers. And the human brain is not naturally suited to think at that level of abstraction. Doing so requires quite a lot of effort and training.

Video Games and Algebra: Some Good News

The kinds of games I have been discussing in this book up to now are not designed to develop any mastery of algebra. They involve no explicit symbolic mathematics, though there can be plenty of algebraic thinking. But that does not mean they cannot have an effect on learning algebra. In fact, you cannot really gain functional understanding of arithmetic without engaging in algebraic-type thinking about generalities across numbers, so any teaching method that develops an understanding of arithmetic must develop algebraic thinking to some extent. Thus, for parents concerned about their children's test scores, the real question they should ask before deciding whether or not to acquire a math ed video game for their kids is: will a student who plays an everyday math video game be better prepared to master (perhaps at school) symbolic math, in particular high-school algebra?

Based on the assumptions many teachers make about learning math, the answer to this question is yes. But is this the case? The usual argument goes like this: algebra involves abstracts across numerical and arithmetical patterns, so the more students master basic arithmetic, and become familiar with numerical patterns and relationships (including functional relationships), the better equipped those students will be to master algebra. For example, the distributive law

$$a \times (b + c) = (a \times b) + (a \times c)$$

is an abstraction across a regular pattern instanced by the following examples:

$$3 \times (7 + 10) = 3 \times 17 = 51 = 21 + 30 = (3 \times 7) + (3 \times 10),$$

$$5 \times (1 + 8) = 5 \times 9 = 45 = 5 + 40 = (5 \times 1) + (5 \times 8).$$

Thus, the argument goes, the more times students are exposed to this numerical pattern, the closer they get to being able to recognize the abstract distributive law, and perhaps even expressing it for themselves.

This sounds reasonable, but there is a problem. It assumes *you can read and understand the algebraic formulas*. But familiarity with, and recognition of, a regular pattern is not the same as capturing it in a single abstract, symbolic expression. This is made clear by the length of time it took the Sumerians to take the step from using distinct counting systems for different kinds of objects to having a single system of abstract numbers. More evidence is provided by the absence of algebraic expressions in the works of the early Babylonians and Egyptians, who instead described their often sophisticated calculation rules in terms of specific numerical instances, both of which I discussed in Chapter 9. And, for a more recent illustration, the subjects in the Recife study and others like it were clearly familiar with many arithmetic patterns, yet unable to handle the most basic symbolic expression. It appears, then, that recognition that a particular symbolic expression represents a regular pattern of numerical instances can come only after the symbolic expression is first understood. This makes mastery of symbolic expressions a decidedly problematic cognitive step.

How does anyone ever make it? I can tell you how I learned algebra, and how I have mastered practically every other new branch of abstract mathematics throughout my career. I read the rules for manipulating the symbols, and I learn them by practicing using them. Sometimes I have a pretty good idea what the rules mean, sometimes only a vague idea, and on occasion they make no sense at all and I'm just baffled as to where it is all going and why. In other words, I learn new abstract mathematics the same way I and everyone else learns to play chess. At first, my use of the rules of a new area of mathematics is like a chess beginner: slavishly mechanical and, to an expert observer, not very smart. But after a while, the rules become second nature, and I no longer think about them. My performance improves, but it's still not good, and I have not yet grasped the "big picture," but I'm better than at the start. Then, after more practice, I become truly fluent, and at that point I find that it all makes sense—I *understand* it.

In the case of chess, the eventual understanding is purely in terms of the game itself, of course. Chess is a one-off game, the rules of which were developed for the sole purpose of making an interesting game. The rules do not have a natural meaning in the everyday world. (Though expert chess players can relate their understanding of chess to real world contests such as wars, battles, and skirmishes—indeed that is how they usually describe their games!)

Mathematics, by contrast, is a vast, interconnected enterprise that is ultimately grounded in the real world. In the case of a newly mastered area of mathematics, the initial understanding is (like chess) in terms of that new area, that is to say, it is a "domain-internal" understanding. But given time and some reflection, the newly mastered material can generally be related to the real world or to other areas of mathematics previously mastered, and provide an understanding of the new material *in terms of things already known and understood*. For instance, a student who has mastered high-school algebra (e.g., being able to make explicit use of the distributive law) will normally have little difficulty seeing how it relates to arithmetic (e.g., how the distributive law applies to particular numbers and can be used in calculation). But going the other way, from arithmetic to algebra, is problematic. The Recife children used instances of the distributive law all the time but could not express or use it symbolically.

Indeed, it seems that prior knowledge of specific examples can sometimes delay mastery of the new abstraction. The problem is that any abstraction captures some but not all of the features of the individual cases. That, after all, is the essence of abstraction. Once you have grasped the abstraction, you can see which features are captured and, more to the point, can recognize which features of any individual case are not pertinent to the abstraction. But until then, there is confusion, or at least the strong potential for confusion. And that confusion can hinder mastery of the new abstraction.

This explains why it usually takes human societies considerable effort and time to introduce and accept new levels of abstraction, particularly in mathematics. On the other hand, once a new abstraction has been introduced, it does not take long before many people master it. This is because of a highly valuable feature of the way the human mind works, exemplified by the way I learn new mathematics that I described a moment ago. Namely, we are able to learn to play initially meaningless games such as chess, to learn to recite intrinsically meaningless prose such as Lewis Carroll's *Jabberwocky*, and to master initially meaningless symbol manipulation rules such as high-school algebra. All it takes to achieve a proficient level of performance is repetition. Understanding is not required. In fact, attempts to understand what it all means at too early a stage can slow the learning process. Practice long enough, however, and reflect on what you are doing, and understanding can (and generally does) eventually emerge. For instance, anyone who is familiar with *Jabberwocky* will have formed a mental image of "the slithy toves" as they "gyre and gimble in the wabe." The human mind cannot stop itself from giving meaning to things, even when there really is none.

The way to learn a branch of symbolic mathematics, such as algebra, is, it seems, to adopt the chess (or *Jabberwocky*) approach. Learn, follow, and practice the rules without trying to understand them. In due course, provided you continually reflect on what you are doing, understanding will generally emerge, particularly if you are given a little help from others. To paraphrase the memorable line from the 1989 Kevin Costner movie *Field of Dreams*, "If you learn it, meaning will come."

So what about our parents' question, "Will a student who plays an everyday math video game (of the kinds I have been discussing) be better prepared to master symbolic math, particularly high-school algebra?" The answer is likely to be "Yes." (We cannot know for sure until such games have been built and released.) Here is why.

As the NRCR made clear, there is more to learning math than the acquisition of skills. In particular, the student's acquisition of a positive disposition toward mathematics is a major factor. And here, I believe video games can have a significant impact. The player who emerges from a video game—with enhanced, real-world, basic mathematical abilities—will, if the game is properly designed, have had a positive, enjoyable, and rewarding experience where solving mathematical problems is a natural and meaningful activity. He or she will, in a nutshell, have developed a productive disposition toward math. This outcome should not be dismissed or viewed as marginal. To my mind, a productive disposition is *the single most important outcome to aim for in K-12 mathematics education.* Get that right, and everything else follows with relative ease. (I did say "relative"; there is no truly easy way to improve mathematical performance.)

Beyond developing productive disposition, if you want to use video games to improve performance in high-school algebra and other areas of abstract, symbolic mathematics, including better performance in written tests, you have to adopt a different strategy in which the video game is just one (albeit crucial) part of a broader framework. Before I describe that strategy, I should make some more general observations about written tests of basic skills.

The Transfer Problem

It is my belief that players who practice mathematical skills in a virtual world will be able to use those skills in comparable real-world situations. That is already a significant achievement; after all, it is in the real world that most people use mathematics! But will they be able to apply those skills on written tests to obtain a better score? As I already indicated, I suspect not, and the reason is the "transfer problem"—something all teachers are familiar with.

Even students who perform well in a particular class and do well on the tests typically have great difficulty applying what they have learned outside the class, either in a real world situation or in another course. The NRCR's broad goal of mathematical proficiency is intended to facilitate transfer, but to date it remains a major cognitive problem. We are hard-wired to learn in meaningful contexts, but what we learn tends to be strongly associated with those contexts. It's just the way we are. That is the transfer problem. In general, only after we learn something in several different contexts and come to recognize the commonalities in what we have been doing do we become able to apply it in a new context, and in some cases we may not reach that stage. Video games can clearly assist in this process; indeed, by being able to present the same mathematical ideas in many different contexts, games seem ideal for the task. But because of the transfer problem, I do not expect that an educational video game will provide a magic bullet for improved test performance.

In fact, to date, only one method has been discovered that works well—brilliantly well, in fact—in improving students' test scores in mathematics: repetitive training geared specifically toward the test, where the students spend hours practicing exactly the kinds of questions that are on the test. Whether this method has any educational value beyond an improved test score depends on several factors, particularly in the case of the kinds of performance tests that seem to be beloved by politicians eager to demonstrate "progress" in education. (I am putting aside benefits such as gaining entry to a particular college. My focus here is on education, not the acquisition of credentials.)

The first factor is that the educational value of doing well on a test depends on what the test asks for. Unless the test focuses on important skills, any value is likely to be minimal, especially in the light of the effort expended. Second, even if the test is well designed to measure important skills, the problem with test cramming techniques is that the mastery they achieve is narrow and brittle. All learning suffers from the transfer problem, and test cramming suffers in spades. For skill drills and skills testing to have lasting value, they have to be accompanied by other educational activities—indeed you need all five threads of the NRCR's mathematical proficiency. This is the context in which we should view contributions made by any new educational initiative, including educational video games.

We will have to wait for the appropriate games to be developed and the inevitable studies of their use to be carried out before we know if I am being unduly pessimistic here, and whether time spent playing math ed video games actually does result in measurable improvement in test taking, either by virtue of improved student disposition, or by building the all-important conceptual understanding of the basic notions of mathematics. I would love to have my pessimism on this issue proved unfounded. There may be some degree of improvement, likely minimal. But I am not holding my breath.

Let me stress again, however, that test taking is just one aspect of learning mathematics. As the NRCR makes clear, broad student performance in mathematics is multi-faceted and depends on many factors. In the case of basic math performed symbolically, it is very important that the students understand the concepts and the methods if they want to get beyond the two extremes of, on the one hand, the Recife children who could perform well in the market but were hopeless at symbolic math and, on the other hand, the school test whiz who knows and can apply all the formulas, but is happy to leave a negative answer (correct apart from the sign error) to a problem to compute an area. Thus, if parents want to look beyond the test at the next educational challenge their children will face, and ultimately the challenge of living happy, productive lives, it is easy to see time spent playing a well-designed math ed video game as yielding value.

How Can You Put Algebra into a Game?

I am proposing that the initial crop of video games designed to develop mathematical thinking should focus on what I believe the medium is most ideally suited for: the development of real-world-applicable mathematical thinking involving everyday mathematics that the learner can make immediate use of in the world. Nevertheless, as I have indicated, I do not discount the importance to future citizens of achieving mastery of algebra in particular. And I am totally aware of the desire of parents to see their children do better at school math tests. I promised I would come back to these issues. Now I will.

The good news is that there is reason to believe that the mathematical thinking and the basic skills developed with the aid of a properly designed math ed video game will have a positive impact on school mathematics performance in algebra, and not just because of the improved productive disposition it engenders in players. In the final report from the presidentially-commissioned National Mathematics Advisory Panel published by the United States Department of Education in 2008, the panel observed that "problems in mathematics learning in the US increase in late middle school before students move into algebra."[2] By addressing

---

[2] U.S. Department of Education, "Final Report of the National Mathematics Advisory Panel," 2008, p. 3.

those prior obstacles to mastery of algebra, a video game can help ensure that students can board the algebra bus, even if there is no guarantee they will arrive at the intended destination. There are a number of things a math ed video game designer can do.

First, although it is not possible to graft symbolic mathematics onto a video game without reducing significantly the quality of the game play (and therefore of the resultant learning), a designer can put in mathematical activities that require algebraic thinking. For example, the solution of simple linear equations can be instantiated in a game in terms of assembling the parts of some simple device. So instead of solving the symbolic expression $3x + 5 = 20$, the player has to determine how many girders of length 3 units it takes to build a simple bridge across a chasm of width 20 units starting from a ledge that stretches out 5 units. To be sure, such an easy example can be solved quickly by inspection or trial and error, but it is not difficult to generate much harder versions where trial and error would either be unfeasible in the game context or else would take a very long time.

Does experience at solving such puzzles prepare the groundwork for a student to subsequently learn the corresponding symbolic algebra? Probably not. But a teacher could take that puzzle as a starting point for a lesson on the algebra behind it, and build on the students' experience and motivation in order to teach them that algebra. The trouble is, as I have observed above, it requires a huge cognitive leap to go from algebraic thinking in specific contexts to mastery of the associated symbolic descriptions and reasoning. To develop mastery of symbolic mathematics, you have to present the student with the symbols. One way is for a teacher to do this, as I just suggested. I'll come back to that idea momentarily.

Another possibility is to take the symbolic mathematics into the game. Although I have been very critical of math ed video games that crudely embed schoolroom symbolic algebra into a game world, there is no reason why you cannot design a world where symbolic mathematics arises naturally. In a fantasy world, there is nothing preventing you from weaving such challenges into the game lore. For example, in order to complete a particular quest in an MMO, a player might have to travel to a wise sage who lives in a cave on a remote mountaintop. Negotiations with the sage necessarily require the player to engage in some symbolic algebra, perhaps in the form of challenges that the sage issues before helping the player complete the quest.

There are many variations on this general theme that can embed some algebra into a game. They are fine as far as they go, and I see no reason to not include such activities in a game. But their inclusion needs to be limited. To make the game work well as a game, such tasks have to be optional. A designer can certainly ensure that performing such a task yields a very desirable reward so that some students may attempt some of those challenges. But if the strategy is to introduce a major topic like algebra through marginal and optional quests in remote locations, not only will many students never do those quests, the game sends the student a message that algebra is yet another ancillary mathematical notion that can be ignored. It is surely far better to handle algebra in a way that demonstrates its ubiquitous power—something that can be used all the time.

To do that, you have to look at the broader environment in which people (children and adults) typically play a video game—the metagame or Gee's big-G game. There I see huge possibilities for learning algebra. The trick is to take what might first appear to be a major obstacle for embedding algebra in a video game and use that to drive the design.

On the face of it, embedding algebra learning in a video game seems impossible. The essence of algebra is to step back from the everyday world, reflect on it—often at length—and think abstractly across many real-world situations, perhaps coming up with a general formula that can be applied repeatedly in different real-world situations.

But the essence of a video game is real-time action in the game world. In fact, if you restrict your attention to what the player's *character* does *within the game* (defined narrowly as the activity inside the game world), the problem surely is insurmountable. While the character can perform many arithmetical tasks, there is no way the character can do algebra. (Except in very special circumstances, such as the wise sage on the mountaintop I speculated about earlier. As I said, it's fine to have those. They just should not be the main vehicles for getting the algebra into the game.) But we are not talking about just a character, rather a player represented by an avatar, and that is a very different identity. The aim, after all, is for the *player* to develop mathematical skills, not the avatar he or she controls. Part of successful game play is for the player to shift, constantly and seamlessly, between the avatar identity and the player identity, to move from seeing the action through the avatar's eyes to seeing it through the player's own eyes. It is by taking advantage of this dual identity that we can provide the player with opportunities to develop algebra skills. Since algebraic thinking is logical thinking abstracted across numerical (i.e., arithmetical) situations, we can, if we are sufficiently skillful in our game design, provide in-game challenges for the avatar so that the most efficient way for the player to direct the avatar to meet them is by the player doing algebra. True, the player must pause in-game action in order to "do the algebra." *But that is the very essence of algebra*—stepping back and reflecting in order to find a higher-order solution that can be applied again and again.

The trick is to not try to force algebraic thinking into the game play—a strategy that, for reasons I have outlined, is surely doomed to fail except under certain game circumstances that of necessity can arise only infrequently. The solution is to set things up so the player *wants* to step outside of the game action and set things up so that in-game success will follow the moment game play is resumed. Why will the player want to stop playing and do something—to do some math, for heavens sake? I'll tell you why. In order to progress in the game, that's why. I am not talking about out-of-game activities that are unrelated to the game play. Quite the contrary; it is all about making progress in the game. The point is that the player will "do the math" (out of the game) precisely to assist his or her character to succeed in the game.

How can a game designer achieve this? The most obvious way, though perhaps not the only one, is through user-generated content. To provide opportunities to learn algebraic thinking, the designer needs to engineer the game so that actions the player performs outside the game (in particular, not through her or his avatar) have a direct impact on the player's in-game character. This facility can be exploited in a number of ways.

For example, suppose the player's character has to perform a series of calculations to crack the combinations of lockboxes found in a cave. The calculations have to be performed several times for a series of similar boxes, using the same procedure but with different numbers for each lockbox, and under significant time pressure. (The increasing presence of mobs, the cave is underwater, etc.) The only way for the player to succeed is to build a small device that can carry out the procedure automatically and rapidly when provided a key input number for each box. To do this, the player leaves the game, figures out and designs such a device—which in essence amounts to formulating an algebraic formula—uploads the device into the game, and when he or she re-enters the game, his or her avatar now has a tool to unlock the boxes. The game designer would provide a template for the construction of such devices, either in a special section of the game or on a game website linked to the game database.

Of course, the player knows that she is working outside the game, motivated by the desire to improve her character's performance in the game. But this is already the dual identity nature of avatar-focused game play that all gamers are familiar with, and it is not entirely different from the way that players use websites such as eBay or Craigslist to arrange trades of game items, or make use of out-of-game websites (such as Thottbot in the case of *World of Warcraft*) to exchange tips and other information in order to complete a quest in the game. Players do this all the time and find it entirely natural. My earlier suggestion that a teacher could use students' experience with my simple bridge-building challenge to motivate and explain linear equations provides another example. The game designer could arrange things so that a bridge has to be constructed in limited time. The only way to succeed is with the aid of the teacher—working out a formula in advance and using it to build and upload a device that will assemble the bridge automatically when the student re-enters the game and embarks on the quest.

With even 3D MMOs now being built to play in web browsers, the distinction between the game and the game website can be blended together, but in cases such as this I think there may be an educational advantage in keeping them separate. Doing the algebra on a separate website carries with it a valuable and power-ful meta-lesson about the purpose and the power of algebraic thinking. The fact that the algebra is done outside of the game (with the game action halted) in or-der to facilitate more efficient subsequent arithmetical actions in the game when game play is resumed, is *exactly* how we want the players to view algebra. Why? Because that is exactly how algebra is used in the real world!

For example, on weekends in the fall, I participate in weekend bicycle races up local mountains. Dan, a racer and the volunteer series organizer, assisted by volunteer helpers (a euphemistic term for spouses), keeps track of everyone's position and time across the finish line, a task sometimes made more difficult because the local police often insist we have staggered starts to avoid 120 cy-clists racing up the hill in a single group. At the end of the series, the results are aggregated and small prizes are awarded based on overall performance. To handle this scorekeeping, Dan wrote a script so his computer could do all the calculations automatically. (He is a computer engineer; this is Silicon Valley, re-member.) Of course, he could have carried out the individual arithmetic calcula-tions in real time at the events, but he didn't. Instead, he wrote the computer program to do it for him. When he wrote that script, he was not playing the role of a bike racer, wearing skintight Lycra, but rather was in his engineering per-sona, sitting in his home office dressed in jeans and a sweater. As an engineer, this particular task was too simple to give him significant professional pleasure. His motivation, and eventual satisfaction, came from using the program in his other world, the world of cycle racing. In short, in order to organize and partici-pate in a series of bike races, Dan stepped out of the world of cycling and solved a math problem—an algebraic abstraction across the arithmetic calculations of a couple of hundred fellow amateur bike racers—in order to use the result-ing computer program within the world of cycling. He has been using that same computer program, with enhancements he has made since he wrote the first version, for several years now. After each weekend's race, the results are usually posted on the web the next morning, with the overall standings for the series up to that date. This is the power of algebra. Step out of the situation and solve a single abstract problem, and then use the solution repeatedly for as many cases (racers and races in my example) as there are.

I deliberately picked an example where an individual steps out of a volun-tary leisure activity to do some algebra to enhance the leisure activity because that is exactly the situation you should be trying to create in order for your play-ers to learn algebra. The pattern is exactly the same in business and industrial applications of algebra—stop the activity, do the algebra, and then use the results when the activity is resumed. It's how algebra really is used in the everyday world. You don't often *see* the algebra being done, of course, since that is typically done "off-line," in private. What you see all the time is the result of that algebra being applied, over and over again, often automatically behind the screen of your calculator, mobile phone, computer, or microwave oven.

In fact, as I have alluded to once already, you see such uses of algebra all the time in the gaming community, where keen video game players such as WoWheads (the in-culture name for *World of Warcraft* aficionados) develop spreadsheets to analyze game play and determine the optimal way to equip the different kinds of characters and formulate strategies to carry out challenging raids. Results are posted on fan sites for others to share. In the case of mainstream entertainment games, this kind of activity is not engineered; it just arises. In designing a math ed video game, the designer should build it to make algebra-oriented activity an integrated part of game play. That's how to use video games to help students learn real algebraic thinking and master symbolic algebra.

As to achieving improved scores on written algebra tests, the best (and almost the only) way to do that is still to practice taking those kinds of tests. Can video games be used to make such practice more palatable? It's possible, but we won't know until we try. One possible approach (and here I'm extrapolating from things I either have tried or at least am reasonably sure will work) would be to design tasks in a fantasy world video game, or in a particular region in such a game, that can be better accomplished only by becoming fluent at answering certain kinds of test questions. Another approach that might be worth trying would be to provide practice tests on the game website, so that performance on the test leads directly (by one of several possible mechanisms) to advancement or the acquisition of assets in the game. I think it would be important that the test items relate in a meaningful way to game content; otherwise taking the test degenerates to simply jumping through hoops. If the video game were going to contribute at all, working on the test problems would have to be self-evident activities that make sense within the game. That would mean designing part or parts of the game around the kinds of questions that will be on the test.

As I said, all of this is pure speculation on my part. But this is just the start. One of the truly exciting aspects of using online video games in mathematics education is that the game database captures every detail of players' in-game activities, and based on what the players do, the game designers—working with mathematics education researchers—can continually modify the game. The original game design has to allow for such modification, but experienced online game developers long ago figured out how to do that. If the approaches I just outlined are tried and prove not to work, the designer can change them until she gets something that does work. Moreover, she can make the changes initially for just one group of students (e.g., on one game server) and compare the results with those obtained with another group of students—something that is extremely difficult to achieve with classroom teaching where there is always the worry that what is really being tested is the quality of the teaching, not the pedagogy.

For example, once a good math ed video game is up and running, the designer could experiment to see what happens when students are presented with timed tests *in the game,* so that the players practice test taking through their avatars. My guess is that this might not work well for all topics, but I could be wrong. (*Timez Attack* seems to work well, and that is nothing more than timed testing of the multiplication tables in a 3D video game wrapper.) The only way to know for sure would be to try it.

Covering Project Work and Advanced Math

The way to make use of video games to help a student learn algebra points the way to the approach a game designer could adopt to handle any other mathematical topics beyond everyday math. Namely, design the game and an associated website so that the learning takes place in the metagame. The game itself thus provides the initial reason for learning and then using the math, which is carried out offline using the game website. The game can provide feedback for successful completion of the math in the form of the player being able to see and use the results of their offline mathematical work, most likely implemented in some device they can make immediate profitable use of.

At an even more ambitious level, the same approach can be used to generate project work in a typical school classroom setting, where the player, or perhaps groups of players working as a team, can go onto the website to design and construct a weapon or vehicle, for example, that their character(s) can use in the game. Whatever the task, the website would provide the framework (and educational scaffolding) required, and do so in a way that enables the results of the players' work to be uploaded and integrated into the game. By designing such activities appropriately, a game designer can put players into a position where the best way to advance in the game is to consult a textbook or ask their math teacher for help, thereby linking with the other learning resources available to the students. In order to link to a textbook, the designer probably needs to pick one or two possible texts before commencing that part of the game design, and make sure that the book really does provide the students with what the game requires from them. If a goal is to encourage the students to ask their teacher for help, the game developer should perhaps provide the teacher with the necessary information, most likely in an access-restricted, teacher's section on the game website.

Though mathematics majors probably have no need of video-game-assisted learning of mathematics, and indeed most would likely shy away from such an approach, the metagame strategy I just described can be used to great effect in the mathematics instruction of university students in other disciplines. For example, Brianno Coller, a mechanical engineering professor at Northern Illinois University, uses a similar video game approach in an advanced mathematics methods course for his engineering design students. The students' task is to design and build a software control system to autonomously drive a racing car round a track, competing against cars controlled by the systems designed by other students in the class. This is a challenging, semester-long project that requires the students to learn a considerable amount of mathematics. The racetrack and race cars, complete with all of the correct physics, are part of a simulation developed for this purpose by the National Science Foundation. The students get to test out their designs (and hence the mathematics they have done as part of those designs) on a regular basis, by actually loading their software into one of the race cars and seeing how it fares. The final examination consists primarily of seeing how well their cars perform in the grand race tournament held at the end of the course. Of course, the students typically want their car to win. But the true measure of the learning is whether it performs well—not crashing, going sufficiently fast, etc.

In an advanced university course, where the professor decides what format to use to evaluate the students' performance, it is possible to use the course video game not only as part of the instruction, but also to grade performance. There is no pedagogic reason why the same could not be done for state testing in the K-12 system. Indeed, I would argue that there are good educational reasons to do just that. It will be interesting to see how long it is before that happens. First, we have to build the requisite video games. That is the next step we all face.

Finally, I want to say something about how teachers could make use of the kinds of video games I have been discussing. Accordingly, the next and final chapter is not so much about how to design math ed video games, but rather the pedagogy for using them once they are available.

 **6+6 A New Pedagogy**

Frequently, when I tell someone I am investigating the use of video games to teach mathematics, their response is to tell me that that no technology can replace a good teacher. It's a peculiar reaction that I think is unique to the United States. Telling an American I am writing a mathematics textbook does not yield the same response, so technology must be the trigger. For some reasons and in some contexts, education among them, the assumption seems to be that the purpose of technology is to replace people. It's particularly perplexing when you consider that today's Americans have homes and workplaces full of technology, all of which we acquire to *help* us. They help us to work more efficiently or more safely, to enjoy our leisure time more, to stay in touch with one another, and to see and experience things we would not otherwise do. So too, technology can improve education. But it's adding something, not replacing anything or anyone.

Let me be perfectly clear about what I am proposing math ed video games should do. They are intended to be new educational resources that enhance the current learning environment of teacher, classroom, textbooks, calculators, personal computers, television, videos, friends, and parents. They are not intended to replace anything—though they will change the way teachers, in particular, organize their classes and spend their time. And there is reason to hope it will result in more parents, grandparents, and siblings becoming involved in a student's educational activities. I have already given some indication in my account of how a video game could be used in the teaching of high-school and college-level mathematics, and I'll say a bit more later in this chapter. But for the present discussion I'll focus once again on middle-school level *everyday mathematics*, where its role is most central.

The Recife Factor

Here is how a teacher might typically use a video game to introduce a class to a new topic. First she checks her teacher's guide to see which video game (or what part or geographic region of a larger game world) covers that topic and which particular challenges in the game may be appropriate for the class. Then, before she says anything about the new topic, she assigns the class to play that video game (or explore that region—or "zone"—of the game world) and see how many challenges they can meet, perhaps listing some particular quests they should try to complete. If the game is well designed, it is highly likely that many or most of the students in the class will have already explored much if not all the game world. So much the better. Ideally, the students should continue until they meet a challenge that beats them. This will occur at different stages for different students. During the course of their exploration, the students will have been exposed to the new mathematical concept or method, and will have had to do some mathematical thinking, though none of it abstract or symbolic (unless they choose to analyze a problem that way). The video game, remember, is built to *instantiate* mathematical concepts and processes.

For example, picking a combination lock might involve finding the least common multiple of three numbers. The first few times the students meet this task, trial and error works, but then they encounter a lock where that approach takes too long. The students are unlikely to have *learned* any new mathematics in a traditional sense. Rather, the game has set the scene for learning, and has provided motivation and meaning—in advance—for what is about to be learned.

The teacher begins her class the next day by noting where each student got stuck in the game (an easy task since the game can provide teachers with all of the

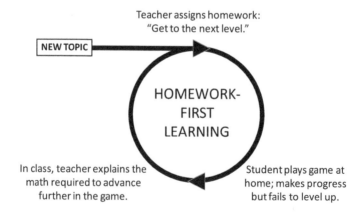

Figure 6. The learning cycle: With video-game assisted learning, exploration (play) always precedes instruction.

statistics about every students' performance the moment they log off), and then guides them in learning the mathematics that will help them meet the challenge when they next re-enter the game. The teacher decides how to do this. It could involve one or any combination of giving a short explanation at the blackboard or asking the students to check out an instructional video, consult the textbook, do a web search, work together in teams, etc. See Figure 6.

When the students subsequently re-enter the game, the game provides an immediate reward for what has been learned, namely being able to open that really tricky combination lock and retrieve the treasure in the chest.

Occasionally, a more extensive video game challenge will require that the student, or a group of students working as a team, enter the game together or go onto the game website to solve a more substantial mathematics problem. For example, it may involve symbolic or geometric reasoning, perhaps to design a tool or device to be used later in the game.

Where exactly is the mathematics being learned? Everywhere. The video game is designed so that learning takes place in the entire system: game, teacher, classroom, family, home, and the game website.

One significant contribution the video game makes is what I have been calling the Recife factor; it gives the mathematics meaning by embedding it in a real context—albeit where that "real context" is within a virtual world and does not have to resemble anything in the real world we live in. When the students first encounter the abstract, symbolic mathematics—most likely in class—they have had an *experience* that provides meaning for those symbols, and they are motivated to master the material. (Again, let me stress that, if you are not familiar with video games, it is hard to appreciate just how compelling and motivating they can be.)

The above discussion indicates that when a video game enters the picture, the role of the teacher should change, from the traditional, somewhat imperial one of "sage on the stage" to the more collaborative "guide on the side." But that change began before people started to talk about using video games in education, and many teachers already embrace the guide model of teaching. If the teacher wants to, she or he can arrange to meet the students in the game (by way of their avatars) at a certain time in the evening to be a guide in the game, answering their questions and assisting them online in the course of the students' game play.

Clearly, this is a very different way to teach mathematics, and some will question whether such change is necessary. Some already have. So why do I believe this change is required? Not because the mathematics has changed. It hasn't. We still multiply 17 by 35 in the same way that Leonardo of Pisa described in 1202, and the way we solve equations today is much the same as the several methods you'll find in Leonardo's book.[1] But the world in which people use that math has

---

[1] See pages 21–22.

changed enormously. Surely, today's world might well require different ways of teaching math. (You might think that allowing the use of the hand calculator is a major change in the way mathematics is taught and done, but not really; Leonardo described a method for performing arithmetic rapidly on your fingers that is accurate for numbers up to 10,000—an impressive "hand calculator" if ever there was one, and moreover one that does not need a new battery every now and then.)

Of course, if the 800-year-old way of teaching math worked, then it would be easy to understand why we still use it. As the saying goes, if something ain't broke, don't try to fix it. But it is broken, and it has been for at least several decades now—at least in the United States. In fact, it did not work that well when the founding fathers were drafting the Constitution, as we learn from writings from the time. And if you ask me, the "golden age of mathematics education" recalled by some commentators, when all students learned how to do math by mastering the basics, probably never existed.

What Has Gone Wrong?

It doesn't take a rocket scientist to identify the problem—which is just as well since rocket science requires a level of mathematical ability that fewer and fewer Americans are achieving. First, mathematics is hard to learn. It is a way of thinking that the human brain finds unnatural. Our brain is a marvel in the animal kingdom, but its great power lies in language, pattern recognition, reasoning by analogy, and the ability to make rapid decisions based on little information. It is particularly ill suited to the methodological, linear reasoning and total precision of mathematics. We have to train our minds to do math, and it's hard work.

Given sufficient motivation to do that work, all but a tiny minority of people can master basic math. (Around 3% have a recognized condition known as *dyscalcula*, which renders math unachievable.) But note my initial caveat about motivation. In the days before the availability of cheap electronic calculators and automated checkout systems in shops, it was important for every citizen to master elementary arithmetic. Everyone could recognize that importance, and by and large everyone did master it—though for most people that was as far as they got. Algebra, in contrast, was unaccompanied by motivation of any likely use, and hence remained beyond them.

For today's Americans, however, there is no obvious, immediate need for any mathematics, so it's not surprising that few see any need to endure the difficult (and, it should be admitted, often tedious and frustrating) process of mastering the subject, especially when there is a whole range of choices of other ways to spend their time that most people find more appealing. And make no mistake about it, much of the mathematics curriculum mandated by politicians and policy makers, and expected at some schools, is not truly needed by many. For example, classical Euclidean geometry is required by most high schools and states for graduation, but it is totally unvalued by most university mathematicians and is a dead topic in most collegiate curricula.

In fact, in today's calculator-drenched world, there are just two kinds of countries where large numbers of children still become proficient in mathematics. One whose educational system we might wish we could emulate; the other we most certainly do not. You can find examples of the first kind by looking at the state education systems of Finland and Singapore, both of which made radical changes to their education system several decades ago that produced remarkable results. I'll take Finland as an example, since, unlike Singapore, Finland has participated (along with the United States) in the Programme for International Student Assessment (PISA), a joint survey of the Organization for Economic Cooperation and Development (OECD)[2] member countries and a number of others. Thus we can compare their performance with ours, though much of what I'll say about the Finnish education system is true in Singapore as well. Finland is a fairly small country, with a population of just over five million, and is perhaps best known today as the home of the Nokia mobile phone empire. (Nokia is a small town in the Finnish forests. The Nokia company began there, manufacturing rubber boots before branching out into the then emerging mobile phone business—a huge shift showing remarkable vision!)

The PISA tests are administered in schools every three years to 15-year-olds, and cover mathematics, science, and reading literacy and problem-solving skills. Tests are typically administered to between 4,500 and 10,000 students in each country. The main focus of the first test (PISA 2000) was reading, the second (PISA 2003) was mathematics, the third (PISA 2006) was science, and the most recent (PISA 2009) was reading again. Six countries have consistently made it into the top ten: Finland, Canada, Japan, Netherlands, Australia, and New Zealand.

In PISA 2003, out of the 30 countries in the OECD, the United States ranked 18th in mathematics, 22nd in science, and 28th in reading literacy and problem solving. In 2006, American students ranked dead last, 25th out of 25, in math and 21st out of 30 in science. In 2009, with 33 countries listed, the United States ranked 25th in mathematics,17th in science, and 14th in reading. Only 2 percent of American 15-year-olds could perform at the highest level, and some 23 percent of American students appeared to have essentially no math skills at all.

The skills of the Finnish students were among the best in all domains assessed in all four PISA surveys. In reading literacy, they placed first in the 2000 and 2003 surveys and second in 2006 and 2009; in mathematics, they rose

---

[2] The Organization for Economic Cooperation and Development, founded in 1961, seeks to bring together the governments of countries committed to democracy and the market economy from around the world to maintain their economic progress. Clearly this is a club the United States wants to be a member of.

steadily from fourth in 2000, to second in 2003, to first in 2006, and now second in 2009; in science, they were third in 2000, then first (jointly with Japan) in 2003, then first in 2006 and 2009. In problem solving skills (not assessed in 2000), they were joint second in 2003, and in 2006 their score was the best result ever achieved in any subject area in any of the PISA surveys. In 2009 they were again second in this category. Clearly, we could do well to ask what Finland is doing that we are not.

Some differences are dramatic. The variation in achievement between the strongest and weakest students in Finland is consistently among the smallest in the surveys. Differences among their schools and regions are also remarkably small. Performance varies only slightly among the different language groups in Finland, and socioeconomic background has a lower impact on students' performance than in the other PISA countries. So, high performance can be achieved with relatively low differences in performance among students.

What is the Finnish secret? It's certainly not time spent in school. On average, Finnish students spend less time per week studying than their counterparts in the OECD countries. Nor is it money spent on education. The Finns' annual expenditure on education is the OECD average. The answer is the way they organize their entire education system, coupled with the way Finnish society views the teaching profession.

In the early 1970s, the Finns completely overhauled their school system, putting in place a system of public education for all children from grades one through nine that teaches in totally unstreamed classes, with students of all academic abilities and socioeconomic backgrounds learning together. Today, teachers for all grades must obtain at least a master's degree, and entry into teacher education programs at universities is highly competitive. Teachers are well paid and enjoy high prestige in Finnish society—comparable to doctors and lawyers. Students start primary school during the year they turn seven, with free preschool as an option before that. Classes average 20 to 25 students. Schools choose their own textbooks, design their own curricula, and allocate their own funds. Since 1998, parents have had wide freedom to select the school to which they send their children.

The features that make the Finland (and Singapore) education systems so successful may be summarized as follows:

1. Teachers are extremely well trained to high national certification standards, including content knowledge.
2. Only the best-qualified applicants are admitted to education school, through a highly competitive process.
3. Teachers are paid a salary comparable to that of an engineer or a doctor.
4. Teaching is a highly regarded profession that many aspire to.

5. Teaching really is a profession, complete with peer development, regular in-service training, etc., and not just a job.

In principle, these features could be made to work in the United States, but most likely they will not. Here is how each feature compares in the United States: (1) some states have tried to demand high certification standards only to find that there are no teachers available, with up to half leaving the profession within five years; (2) fewer and fewer students are entering the teaching profession and typically there is little or no encouragement in academic departments on campus for teachers at any level, including graduate school; (3) teachers' salaries are very low compared with those of other professions; (4) teaching is not always valued and there is almost no encouragement for those entering the profession; and (5) professional development is required but is rarely academically oriented toward content; rather, social issues have taken precedent in many schools.

The other dynamic that produces well-trained mathematicians, the one we surely do not want to emulate, can be found in China and India. In those countries, there is enormous pressure on children to do well in math and science, because success in those subjects offers a ticket to a far better quality of life than would otherwise be possible. This is very much a survival of the fittest model, fueled by both self-motivation and parental pressure, with thousands falling by the mathematical wayside for every one that truly makes it. You could find a similar (though perhaps not as intense) highly competitive atmosphere in Europe and to a somewhat lesser extent the United States in the first few decades after World War II, but that dynamic has long since been eradicated as the western world became more affluent.

Without the presence of known drivers that produce citizens who are competent in basic math, what can a country like the United States do? Well, the first thing we can do is admit that the existing approach simply does not work for us, and there is no way it can be made to work. Then we need to step back and take a fresh look at the situation. What exactly do we want to occur and why?

Today's Need: The Innovative Mathematical Thinker

I'll admit that I would like to see us adopt an education system like the ones of Finland or Singapore, having the features I listed above. But even if we could—and as I already indicated, I do not think it would be possible in the United States—I don't think it would be enough anymore. The world has changed. In the words of Thomas Friedman, today's world is flat.[3] Focusing entirely on the business and

---

[3] Thomas L. Friedman, *The World Is Flat: A Brief History of the Twenty-First Century*, Farrar, Straus and Giroux, New York, 2005.

commercial worlds, traditional mathematical tasks can be outsourced even more easily than manufacturing, at electron speed with virtually no shipping cost. A designer in San Francisco who wants a certain set of equations solved can simply email them to a specialist center in India and the answer will be emailed back by the next morning. Faced with that economic reality, the only viable response for the United States is to do what it has been doing for the past century, and what we have already done with manufacturing, which is to stay ahead of the curve as the world's main innovation engine. Just how long this strategy will work is hard to predict—the future usually is—but I don't think we have a better choice.

For many years, we have grown accustomed to the fact that advancement in an industrial society required a workforce that has mathematical skills. But if you look more closely, those skills fall into two categories. The first category comprises people who can take a new problem, say in manufacturing, identify and describe key features of the problem mathematically, and use that mathematical description to analyze the problem in a precise fashion. The second category comprises people who, given a mathematical problem (i.e., a problem already formulated in mathematical terms), can find its mathematical solution. Until now, our mathematics education process has focused primarily on producing people of the second variety. As it turned out, some of those people always turned out to be good at the first kind of activities as well, and as a nation we did very well. But in today's world, and more so in tomorrow's, with a growing supply of the second type of mathematical people in other countries—a supply that will soon outnumber our own by an order of magnitude—our only viable strategy is to focus on the first kind of ability, and hope we can hold our own in that category.

In other words, the only mathematical niche I can see for the United States—and fortunately for us it is a crucial niche in today's world economy—is at the innovation end. Innovation is an area where we still lead the world, in large part because our political system allows and rewards innovation, and also because it is very much a part of the American character. Incidentally, in case that last statement comes across as being American-centric, note that it surely is simply a reflection of our history. For several centuries, and particularly the last century, some of the most innovative people from around the world have flocked to our shores to make their fortune—or in many cases simply to survive. We have a culture of, and a liking for, innovation because that was one of the consequences of large-scale immigration. (We took in many of the world's innovators, taking advantage of the fact that one country's troublemaker can be another's innovator!) Traditionally, a mathematician had to acquire mastery of a range of mathematical techniques, and be able to work alone for long periods, deeply focused on a specific mathematical problem. Doubtless there will continue to be native-born Americans who are attracted to that activity, and our education system should support them. We definitely need such individuals. But our future lies elsewhere, in producing mathematical thinkers that fall into my first category above; what I propose to call the *innovative mathematical thinkers*.

In order to exhibit the abilities I attributed to this category, this new breed of individuals (well, it's not new, I just don't think anyone has shone a spotlight on those people before) will need to have, above all else, a good conceptual (i.e., functional) understanding of mathematics, its power, its scope, when and how it can be applied, and its limitations. They will also have to have a solid mastery of basic mathematical skills, but it does not have to be stellar. (That's just as well, as I was coming close to defining a near impossible ideal.) A far more important requirement is that they can work well in teams, often cross-disciplinary teams, they can see things in new ways, they can quickly come up to speed on a new technique that seems to be required, and they are very good at adapting old methods to new situations.

It should go without saying that possibly the worst way to educate such individuals is to force them through a traditional mathematics curriculum, with students working alone through a linear sequence of discrete mathematical topics. To produce the twenty-first century innovative mathematical thinker, you need project-based, group learning in which teams of students are presented with realistic problems that will require mathematical and other kinds of thinking for their solution.

The (New) Role of the Teacher

In the educational scenario I am describing, the teacher's role is much more important than in the traditional system. In fact, technology has now rendered obsolete much of what teachers used to do. Except on rare occasions, today's teacher should not be spending much time at the board explaining basic techniques. That form of instruction can now be found in videos and interactive instructional materials on the web—much of it for free—where the student can proceed at her or his own pace, free from unwanted distractions by other students, and can stop the video at any moment, view a single frame for as long as required, and replay a segment as many times as necessary. Moreover, such materials are increasingly being offered in a package that tracks each student's progress and delivers new material at a rate that matches the student's ability and past performance. (The non-profit Kahn Academy is an example of such a resource that has garnered a lot of attention of late.)

This role frees the teacher from being a very expensive delivery system that ships facts from textbooks into students' notebooks (and hopefully into the students' minds as well), to being a full-service *learning resource*. Teachers can diagnose what students understand already, including their typical misunderstandings, they can offer alternative representations, counter examples, and examples to help make things clearer when they have had feedback from students. Teachers can explain misconceptions, and they can put students in situations where they have to pay attention to a new idea. I am not saying the teacher will never stand at the board and provide instruction. A well-designed class lesson can be extremely useful. For example, some of the best designed lessons in Japan, China, and Singapore can be very effective because they draw on what students already know and anticipate plausible misconceptions that can arise.

But overall, modern technology changes the balance between the different activities dramatically. In the traditional approach, students were supposed to acquire new facts and learn new techniques in class, and then practice them at home. In the twenty-first century educational system, these two can be, and in my view should be, almost completely interchanged. Home should be the primary location where the student acquires the facts and learns the methods (mostly from the web), and the classroom should be where, with the help of other students and the all-important teacher, the student works through exercises to gain understanding, with the teacher engaging in all the activities I listed a moment ago.

With class projects driving the entire process, it is quite likely that different students will access different material at different times, in different orders, as each team tries its own approach. In the traditional educational model, no teacher could handle that kind of information delivery load, but web videos take care of that with ease. The "facts of math" are few. What takes time and effort is learning how to make good use of those facts—*learning how to think mathematically*—and that is where there is no substitute for a good (human) teacher. That's what today's teacher should focus her or his time on.

Nothing that I said here has anything to do with bringing video game technology into the educational process. The changes in pedagogy that I have described were long overdue when they started in a few schools, and the sooner they have become commonplace the better our educational system will be. What video games can do is greatly enhance this new form of education. Moreover, they can do so in a manner that is particularly well suited to the United States and the other highly developed countries of the world. I get attacked by traditionalists whenever I say it, but in today's developed world—full of things that attract all of our attention—math teachers have to "sell" their subject to their students. Yes, those students are a captive audience, but unless the teacher can motivate them and inspire them to want to learn mathematics, they will do little more than go mindlessly through the motions, and even if they manage to pass the exam, they won't learn much math.

For a few students, the ones who see mathematics as challenging fun from the start, there is no problem, of course. But for most students, arousing their interest comes down to making it relevant to their lives and their interests, or to piquing their curiosity from time to time. For example, in giving talks to middle-

and high-school students over the years, I've found any of the following topics arouses interest:[4]

- Why do golf balls have dimples? (It makes them fly up to 2.5 times farther.)
- If the dimples on golf balls make the ball fly farther, why don't airplanes have dimples? (It turns out that some do have the mathematical equivalent of dimples.)
- What really keeps an airplane in the sky? (The answer you find in most math books[5]—and many pilot instruction manuals—is wrong.)
- Why do honeycombs have a regular hexagon pattern, like a tiled bathroom floor? (The answer is efficiency, but it took mathematicians 2000 years to figure it out correctly. The bees were faster.)
- How many different kinds of soap bubbles can there be? (The mathematics shows that there are some that are theoretically possible, but they are too unstable to survive long enough for us to ever see them, except on a computer screen.)
- How do birds and fish find their way when they migrate between the seasons? (It turns out they are naturally evolved trigonometers.)
- Where does a skateboard get the vertical upward force to leave the ground when its rider executes a jump?
- Why does the supposedly random Shuffle mechanism on the iPod often give you some songs much more frequently than others? (The answer is because it is random.)
- How many different ways can you repeat a design to create wallpaper? (The answer is 17.)
- Is it true that the ancient Greeks used the golden ratio to design their buildings, including the Parthenon? (No, it's a myth, along with many other oft-repeated claims about the golden ratio in art, music, and even in human physiology.)[6]

---

[4] This list is taken from my article "Bringing Cool into School," to be published in the *NCTM Yearbook 2011*. I have also used them as motivating examples in various articles and books I have written, among them *The Math Instinct* (Thunders Mouth Press, 2005), *Life by the Numbers* (John Wiley, 1998), *The Language of Mathematics: Making the Invisible Visible* (W. H. Freeman, 1998), and *The Numbers Behind NUMB3RS: Solving Crimes with Mathematics*, co-written with Gary Lorden (Penguin-Plume, 2007).

[5] Including my own book *The Language of Mathematics*, cited in the previous footnote!

[6] This bizarre belief, for which no one has ever produced a shred of evidence, and which historians have traced back only to the nineteenth century, lives on, though it should have been put to rest by the evidence provided in the excellent scholarly article "Misconceptions about the Golden Ratio," by George Markowsky, published in *The College Mathematics Journal*, Vol. 23, No. 1 (January 1992), pp. 2–19, and elaborated on at even greater length by Mario Livio in his book *The Golden Ratio: The Story of PHI, the World's Most Astonishing Number*, Broadway Books, 2003.

- Why do the famous Fibonacci numbers keep turning up in the natural world? (It's another example of nature being efficient, and the mathematics is really neat. This is one case where the golden ratio really is involved.)[7]
- Does a baseball outfielder run to catch a ball the same way a dog does? (It turns out they both use the peculiarities of the visual system to take advantage of the same, pretty complex, math.)
- Where did cycling legend Lance Armstrong find up to eight minutes by which he won each of his seven Tour De France victories? (It wasn't drugs, unless you think math and science are drugs.)[8]
- Speaking of Lance, just how does a bicycle turn? (The mathematics of this one is much more complicated than you ever imagined. You don't turn a bicycle by turning the handlebar the way you turn a car using the steering wheel. If you try that you'll fall down. You do one of two things; you either keep the handlebars straight and simply lean to the side you want to turn, or—and this is the one most of us do, albeit without realizing it—you turn the handlebars the opposite way to the direction you want to turn, a maneuver called countersteering. This is counterintuitive. You need to think mathematically to understand why things work this way.)

It has to be said that in many of these examples—and they are just a few examples—the mathematics is well beyond the ability of my audience. Indeed, some are beyond *my* ability. But being able to *do* that math is not what is required, of either the teacher or the student. The examples are intended to arouse interest and to demonstrate the relevance of mathematics to everyday aspects of modern life.

The need to arouse interest and motivate students is particularly acute in the case of middle-school mathematics. Though mental arithmetic was an important skill when I was growing up, it isn't today, which means that most of the middle-school mathematics curriculum is actually not of direct relevance to our students' lives. (At least not in any way they will readily recognize. You could try to make a case for the importance of quantitative literacy to good citizenry, but with a 12-year-old you're probably fighting a losing battle if that is your approach.) Of course, the age-old argument that middle-school mathematics forms the basis for more advanced math learned later—math that is relevant to their lives—is a valid one, but again in my experience it does not get you very far with most pre-teens (or even teens for that matter). This is where video games can make what I think is their greatest contribution. In a fantasy world created in a computer, you can

---

[7] See the Livio book referenced in the previous footnote.
[8] There is an excellent Discovery Channel documentary on this called *The Science of Lance Armstrong*.

gy*185

give "real world" relevance and importance to any mathematics you want. That's the Recife factor at work.

Yes, but What about the Test?

Pedagogy aside, parents, teachers, school authorities, and in due course politicians, will surely want to know the answer to this question: can the use of video games help raise the level of middle-school mathematics performance across the board so that we are among the world leaders? Before I can answer this question, we need to agree on that it means. In particular, what does it mean to be a "world leader"?

In terms of sheer numbers, which is the statistic of most interest to the leaders of large corporations who are faced with hiring enough mathematically qualified employees, we cannot hope to keep up with China (population of 1.3 billion) or India (population of 1.1 billion). As I noted before, in both countries there is enormous pressure on children to secure a good education leading to a secure future, both parental and self-motivational, and that will inevitably produce more and more highly able mathematicians, scientists, and engineers. Will the Unites States, with a total population of 300 million (less than a third of each of those two giants), produce individuals equal to those from China and India? Of course we will; no nation has all the world's human talent. But in terms of the overall numbers, there is no way we will be able to keep up.

Or does being a "world leader" mean that United States students will rank number one in international rankings such as PISA or TIMMS, which are the statistics that politicians seem to care most about? I seriously doubt it, and if you have read this far you will have realized by now that I don't think this should be our main goal. Even if it were, the hard reality is that currently we don't have the educational system or the societal values to be able to do what, say, Finland and Singapore did, and transform our educational systems so that we are at the top of the rankings. Nor do I see any signs of a political or social will to change things in the way those (much smaller and much more cohesive) countries did. You can, of course, lament this state of affairs all you want, but that is the playing field on which those of us with a stake in United States education have to operate. So let's work with what we've got. As one of the most innovative and resourceful nations on Earth, we might just find that in the process, we find ourselves once more in a leadership role.

That leadership role is to be found in the growing strategic importance (particularly to the Unites States) of *innovative mathematical thinking*. This is where our economic future lies as a nation. And that is what our educational system needs to embrace. It is an educational philosophy that our local-control, bottom-up, free enterprise, federated social and governmental structures, and, let me add, our national character are particularly well suited to. Well-designed video games can play a major role in developing the kind of free thinking, novel ways of thinking about problems, and making use of mathematics when required or appropriate, that are necessary in a nation whose economic well being depends upon constantly innovating.

The business world, always eons ahead of the educational community, realized this long ago. In their excellent, informative, and thought-provoking book *Total Engagement: Using Games and Virtual Worlds to Change the Way People Work and Businesses Compete*,[9] Stanford professor Byron Reeves and entrepreneur and venture capitalist J. Leighton Read describe how some of the world's most successful multinational companies already use MMOs to develop innovative thinking among their key employees (as well as take account of experience in games such as *World of Warcraft* in making hiring decisions).

Once we have a number of video games of the kind I am advocating, and teachers have become familiar working with them (today's students will take to them immediately—from their perspective they will be just more video games), a large part of the end-of-year student achievement test could take place in the video game, or, for some test questions, using a combination of the video game, the textbook, and the game website. Maybe the entire test could be given that way. Only time and experience with such games would let us know for sure.

Or maybe society will still feel a need for some written tests. In any event, written tests are not likely to go away any day soon, so in designing math ed video games we must address the issue of how their use will affect student test scores in current testing regimes. Again, there is no way we can be sure without actually building some and putting them into use. But since they will be an *addition* to the current system, it hardly seems likely that test scores would fall! Students will, after all, still be spending a lot of time doing paper-and-pencil, symbolic mathematics. In fact, I would expect test scores to rise. All the available evidence strongly suggests that video games will at the very least improve both students' attitudes toward mathematics and their ability to use mathematics in "real world" contexts. And that will mean that they approach that paper-and-pencil symbolic stuff with a much more positive attitude than is currently the case. You are not convinced. You want more evidence? Well, let me give you some. Meet Jo Boaler.

Jo Boaler

Jo Boaler[10] is a Professor of Mathematics Education at Stanford University. She began her career as a math teacher in her native United Kingdom, transferred

---

[9] Harvard Business School Press, 2009.

[10] Most of this section first appeared in my monthly column "Devlin's Angle" on the website of the Mathematical Association of America (www.maa.org) in June 2010.

to academia (London University), and was for several years a professor of mathematics education (and thus a colleague of mine) at Stanford. Then, in 2006, she returned to the UK to take up the newly established Marie Curie Chair in Mathematics Education at the University of Sussex in England. In 2010, she returned to Stanford. Boaler's research focuses on what kinds of teaching are more effective. But she has taken a somewhat unusual approach. She surveyed the students both while they were at school and then some years after they had graduated and entered the workforce. Some of her findings are truly illuminating. Over many years, Boaler has conducted interviews with hundreds of students from both traditionally taught math classes (where the teacher explains a method at the board and works through a few examples, and then gets the students to do a number of similar problems) and those with a more progressive approach (project-based, collaborative team work). One of the questions she asked them was what it took to be successful in math. By far the most common answer she received from students taught in a traditional fashion was to *pay careful attention*.

Among other answers Boaler received in schools with a traditional pedagogy, which she quotes in her recent book *What's Math Got To Do With It?*,[11] are:

"I'm just not interested in, first, you give me a formula, I'm supposed to memorize the answer, apply it, and that's it."

"You have to be willing to accept that sometimes things don't look like— they don't see that you should do them. Like they have a point. But you have to accept them." (p. 41)

Another traditionally taught student who Boaler interviewed was conscientious, motivated, and smart, and regularly attained A+ grades in mathematics. She was able to follow the methods her teacher demonstrated in class, and could reproduce them perfectly. But she did not understand what she was doing, and as a result she regarded herself as not good at math. When Boaler asked her why she thought that, she replied, "Because I can't remember things well and there is so much to remember" (p. 164).

Over a four-year period, Boaler followed the progress of 700 students through their high-school careers at three high schools. One of the three was in an urban setting, close to a railway line, which (for human-subjects research confidentiality regulations) she refers to in her book by the fictitious name "Railside High." She first visited the school in 1999, having heard that they seemed to be achieving remarkable results, despite the poor location and run-down appearance of the school buildings.

---

[11] Penguin Books, 2009.

Boaler tells us that a number of features singled out Railside as her school to study. First, the students were completely untracked, with everyone taking algebra as their first course, not just the higher achieving students. Second, instead of teaching a series of methods, such as factoring polynomials or solving inequalities, the school organized the curriculum around larger themes, such as "What is a linear function?" The students learned to make use of different kinds of representations, words, diagrams, tables, symbols, objects, and graphs. They worked together in mixed ability groups, with higher achievers collaborating with lower performers, and they were expected and encouraged to explain their work to one another. (pp. 58–68)

Parents whose own math education was more traditional, with the students sitting in rows, in ability-streamed classes, being shown methods by the teacher and then working silently on their own—and that is practically all parents—often find it had to believe that the Railside approach could work. They believe the loose structure would mean that the kids won't master skills well enough to pass tests, and that the presence of weaker students will drag down the better ones. Often they maintain this belief despite freely admitting that the traditional approach did not work for them and is contrary to their own experiences every day at work where over many years they have discovered that collaborative working is highly effective, and that when someone who knows how to do something assists someone who does not, *both* learn and benefit from the experience.

In the nineteenth century and for much of the twentieth, most industrial workers *did* work silently on their own, in large open offices or on production lines, and under the supervision of a manager. Schools, which have always been designed to prepare children for life as adults, were structured similarly. An important life lesson was to be able to follow rules and think *inside* the box. But today's world is very different—at least for those of us living in highly developed societies. Companies long ago adopted new, more collaborative ways of working, where creative problem solving is the key to success—thinking *outside* the box. Companies that did not adapt went out of business, but by and large the schools have not yet realized they need to change and start to operate in a similar fashion.

Of course, it may, as many parents seem to assume, be different for children in schools. After all, they will argue, what works for adults may not be successful for children. That's a fair concern. It's a concern that is addressed head on by Boaler's findings. The other two schools Boaler studied along with Railside were in more affluent suburban settings, and the students started out with higher mathematics achievements than did those at the urban Railside school. Since those two schools adopted a traditional form of instruction, Boaler was able to compare student outcomes over the entire four years of high school. By the end of the first year, she found that the Railside students were achieving at the same levels as the suburban students on tests of algebra. By the end of the second year, the Railside students were outperforming their counterparts in the two suburban schools in both algebra and geometry tests. By their senior year, 41 percent of Railside students were in advanced classes of precalculus and calculus, compared to only 23 percent of students from the other two schools in more affluent neighborhoods.

What's more, the Railside students learned to enjoy math, and saw it as useful. When Boaler and her team interviewed 105 students (mostly seniors) about their future plans, 95 percent of the students from the two suburban school said they did not intend to pursue mathematics as a subject any further, even those who had been successful. At Railside, 39 percent said they planned to take further math courses.

When Boaler would visit a class being taught in a Railside-like fashion and ask students what they were working on, they would describe the problem and how they were trying to solve it. When she asked the same questions of students being taught the traditional way, they would generally tell them what page of the book they were on. When she asked them, "But what are you actually doing?" they would answer, "Oh, I'm doing number 3." (p. 98)

Prior to moving to Stanford, while she was still working in her native UK, Boaler had begun a similar longitudinal study, comparing two very different schools that she called Phoenix Park (in a working class area) and Amber Hill (in a more affluent neighborhood). The former adopted a collaborative, project-based approach similar to that at Railside, and the latter a more traditional pedagogy. (pp. 69–83) Boaler had chosen these two schools because, despite being in different socioeconomic regions, their student intakes were demographically very similar, their entering students at age 13 had all experienced the same educational approach, and the teachers at both schools were well qualified. One difference between the English schools and those in California is that the UK does not follow the US practice of dividing mathematics into separate sub-subjects, such as Algebra I, Algebra II, or Geometry; rather they just learn math (or "maths" as it's called in Britain). But other than that, this was very much like the study she would subsequently conduct in California, and the results were strikingly similar.

At Phoenix Park, students were given considerable freedom in math classes. They were usually given choices between different projects to work on, and they were encouraged to decide the nature and direction of their work. One student explained to Boaler how they worked in these words:

"We're usually set a task first and we're taught the skills needed to do the task, and then we get on with the task and we ask the teacher for help."

Another described the process like this:

"You're just set the task and then you go about it ... you explore the different things, and they help you in doing that ... so different skills are sort of
tailored to different tasks." (p. 70)

In one task Boaler describes, the students were simply told that a certain
object had volume 216, and asked to describe what it might look like. In another,
the students were told that a farmer had 36-meter-long lengths of fencing and
asked to find the largest area the fences could enclose. If you think either of these
is "shallow" or "not real math" then almost certainly you are living, walking proof
that traditional math instruction deadens the mind to see the many possibilities
each task offers, and the amount of mathematical thinking required to carry out
the investigations. In her book, Boaler sketches some of the creative thinking the
Phoenix Park students brought into the two tasks, and the mathematics learning
that resulted. To my mind, what she describes is the early development of the creative, collaborative, problem solving skills that are essential in today's world. As
one student explained to her:

"If you find a rule or a method, you try to adapt it to other things." (p. 74)

While the Phoenix Park students were discovering that math is challenging
and fun, and provides an excellent outlet for their natural human curiosity, things
were going very differently over at Amber Hill. There, the students worked hard,
but most of them disliked mathematics. They came to believe that math was a subject that only involved memorizing rules and procedures. As one student put it:

"In maths, there's a certain formula to get to, say from A to B, and there's
no other way to get it. Or maybe there is, but you've got to remember the
formula, you've got to remember it." (p. 75)

It was at Amber Hill that a student provided Boaler with what I find one of the
most revealing and disturbing student quotations I've ever come across:

"In math you have to remember; in other subjects you can think about
it." (p. 76)

Though the Amber Hill students spent more time-on-task than their
counterparts at Phoenix Park did, they thought math was a set of rules to be
memorized. The ones that were successful did so not by understanding the
mathematical ideas but by learning to follow cues. The biggest cues telling
them how to solve a problem were the method the teacher had just explained

on the board, or the worked example that immediately preceded it in the textbook. Another cue was to use all the information provided in the question, but nothing else. That strategy can be made to work well until the examination at the end of the year, when those cues are not present. Predictably, even the Amber Hill students who did well during the term did poorly in those exams. And, in the national exams that all British students take at age 16, the Phoenix Park students easily outperformed them. Faced with a problem they did not recognize as being of a familiar type, an Amber Hill student might freeze, or struggle in vain to remember the right formula, whereas Phoenix Park students tried to make sense of it, and adapt a method they thought could be made to work.

In addition to her classroom studies at the two schools, Boaler also interviewed the students about their use of mathematics out of school. By then, many of them had weekend jobs. All forty of the Amber Hill students she interviewed declared they would never, ever make use of their school-learned methods in any situation outside school. To them, what they had been taught in the math class was a strange sort of code that can be used in only one place: the math classroom. In contrast, the Phoenix Park students were confident they would make use of the methods they had learned at school, and they gave her examples of how they had already made use of their school-learned math in their weekend jobs.

In a follow up study she conducted some years later, Boaler surveyed the then 24-year-old graduates from Phoenix Park and Amber Hill. When they had been at school, their social class, as determined by their parents' jobs, was the same at both schools. But eight years later, the young adults from Phoenix Park were working in more highly skilled or professional jobs than the Amber Hill adults. Demonstrating how good education can lead to upward social mobility, 65 percent of the Phoenix Park adults were in jobs more professional than their parents were, compared with 23 percent of Amber Hill adults. In fact, 52 percent of Amber Hill adults were in *less* professional jobs than their parents, compared with only 15 percent of the Phoenix Park graduates. (Remember, although the schools were located in different social areas, their student bodies were very similar.) (pp. 80–83)

Of course, you won't get this information from reading the computer-generated scores from standardized tests. Boaler does not find her data by gazing at a computer screen. She goes out and talks to the people who the education is all about: the students and former students. I ask you, which is the more important information: the score on a standardized, written test taken at the end of an educational episode, or the effect that educational episode had on the individual concerned? As a parent (if you are one), which of the following two statements would give you more pleasure?

- "Because of good teaching, my child scored 79% on her last math test," or
- "Because of good teaching, my child has a much better job and leads a far more interesting and rewarding life than I do."

Now throw video games into the mix. Exciting prospect, don't you think?

High-School Math and Beyond

When a student gets to algebra, the nature of mathematics changes. Numbers, though abstractions, refer to things in the world: sizes of collections, the weight of your body, lengths, areas, and volumes of objects, etc. In algebra, however, the abstractions are abstractions from abstractions. (The very sentence already causes us to do a double take before we understand it.) Those x's, y's, and z's generally denote arbitrary numbers. And that's in high-school algebra; in college algebra, the letters can denote all manner of other abstract entities. Because algebra is at least two steps abstracted from reality, it is all but impossible to instantiate it in a physical scenario, even in an imaginary digital world. (You can do it with the simplest cases, such as linear equations and inequalities, but beyond that it's hopeless.) That means it is not possible to provide students with "physical" activities to prepare them for the symbolic representations that follow, the way you can for arithmetic. Which is unfortunate, since many students who do just fine in arithmetic flounder when they meet algebra.

For many of them, the problem is, as I noted earlier, that they do not realize that algebraic thinking is very different from arithmetic thinking. To use my favorite example, setting up a computer spreadsheet is a very different activity from doing arithmetic. In fact, the whole point about spreadsheets is that they do all the arithmetic for you! Since setting up a spreadsheet can be hard work, it's clear then that it is not the same as doing arithmetic. A mathematician would explain the difference by saying that setting up the spreadsheet involves algebraic thinking, not arithmetic.

There already is one obvious strategy for using video games to develop algebraic thinking skills. Pick any popular (entertainment) game on the market and you will find at least one fan site where players discuss performance and suggest strategies for playing the game. And on those sites you will often see spreadsheets that players have posted for analyzing game plays. This occurs even though the designers of those games did not set out to encourage players to develop their algebra skills. Imagine what could be achieved with a video game *designed* so that players gain considerable benefit from carrying out such offline analyses.

More than that, as I indicated before, the designer can put difficult versions of challenges into the game that can be solved only by the player exiting the game, working out a key formula or solving a particular equation—perhaps with the help of the teacher or a classmate—and then making use of that solution when they next log on. Players of many video games are very familiar with constantly coming out of the game to perform some game-related activity that will enable them to advance within the game. They come to distinguish between the two kinds of activity—in some respects it is a distinction between strategic and operational action. But this is precisely why we have algebra and why it is so useful. Consequently, not only will the student who learns algebra in and with such a video game actually do the algebra (offline, most likely with a pencil and paper), and hopefully get better at it, he or she also learns why we have algebra, what it can do that arithmetic cannot, and how, when you solve a problem in algebra, the solution applies to every particular case involving specific numbers.

It is amusing to note that, one consequence of what I just said about leaving the game world to do algebra is that the virtual world of the video game is where the student experiences "reality" and the real world in which the student is physically situated is where she or he abstracts from that "reality." This total swapping around of the real and the virtual is actually not uncommon where virtual worlds and video games are concerned. A friend of mine once remarked that one Christmas her daughter asked for a virtual sword that was available in a fantasy video game she liked to play. The item was particularly hard to find, but could be purchased from online dealers. As it happened, when they were Christmas shopping, the parents spotted a physical model of the sword in a games shop, and thinking that a real sword the daughter could hold in her hand would be preferable to some code in a computer game database, that is what they bought her. When Christmas Day came, the daughter could not hide her obvious disappointment when she saw the present. When asked what was wrong, she replied, "But I wanted the *real* sword." For the daughter, the sword in the game world was real, the one she had been given just a toy copy.

For mathematical topics beyond algebra, the connection to "physical things" in the game world is even more remote. In those contexts, the main educational contributions of a math ed video game are (1) to provide a natural (within the game world) context in which a particular piece of mathematics becomes necessary, and (2) to enable the students to test their mathematics by putting the results back into the game and seeing if things work the way they should. This is the approach Briano Coller takes with his university engineering students. (See pages 171–172.)

Few people, it seems, find doing mathematics intrinsically interesting. But most people like to explore things and to build things. With a video game, we can embed the need for doing the math in intellectually appealing wrappers. Of course, the students will know what is going on, just as Coller's students know they are enrolled in a mathematical methods course. We are not trying to fool them into doing math. But that does not mean this approach does not work. Context makes a huge difference.

Remember, we humans survived as a species because we evolved to be problem solvers. It's in our genes to solve problems, and (as a result of the process of natural selection that put that capacity there) it gives us enormous pleasure to do so. But the problems that we instinctively like to solve almost always arise in the real world, as they did throughout evolutionary history. Mathematics educators have known that since the dawn of the subject. That is why mathematics instruction books throughout the ages have been chock full of word problems. Unfortunately, as every schoolchild quickly learns, word problems in a book don't really cut it when it comes to evoking tingles of pleasure—at least for the majority of students. They just do not offer enough context—a world in which to act. In a video game, it's very different. Once you are cognitively inside that world, it is your reality. As a result, 200,000 years of evolution of the human brain come into play.

Of course, there is one group of students for whom learning advanced mathematics with the aid of a video game will not offer much advantage, if any. Indeed, those students will likely shun math ed video games altogether, just as they shun college math textbooks. Those are the future mathematicians of the world. People like me. We discover very early in our lives that there is a virtual world far more intricate and exciting than any video game will ever be. That world is Mathematics itself. We play in the most fantastic video game there is. But for everyone else, well-designed digital video games will represent a major step forward in mathematics education. For middle-school students, it will be a huge leap forward.

Other Kinds of Video Games

Until now, my focus has been on video games where the action takes place in a virtual world having a history and a purpose. That world can be a fairly simple two-dimensional one; a rich, immersive, 3D virtual world of an MMO like *World of Warcraft*; or something in between. The player is represented in the world by an avatar that he or she controls. There is some sort of back story (varying from extensive to very minimal) that provides players with a history and an understanding of their avatars' characteristics, together with an overriding purpose for the avatar's actions. By making skillful use of such features, game designers and educators working collaboratively can arrange for situated learning to occur during the course of normal game play.

But there is another class of video games that do not have all of those features, the so-called *casual games*. The essence of casual games is they are just that: players take them and leave them. Many are really just interactive puzzles, often requiring rapid responses to visual stimuli. My current favorites are *Bejeweled*, *Bubble Jump*, and *Angry Birds*. There are a number of such games that purport to teach some mathematics, but for the most part they do little besides provide repetitive practice of basic skills under time pressure. I see no reason why there would not be games of this kind that develop mathematical thinking, though I suspect they would resemble *Sudoku* more than *Bejeweled*. The popularity of online *Sudoku* shows that intellectually demanding casual puzzle games can work, but the fact that there has so far been no serious competitor to that puzzle suggests it may be fiendishly difficult if not impossible to cover much of the middle-school mathematics curriculum in a similar way. But we won't know for sure until people try it.

One challenge in designing a casual game to teach mathematics is providing the player with the motivation to persevere. The casual games industry has discovered one way to encourage players to keep playing, at least for online games: human social instincts and peer pressure. A new genre of products called social network games take advantage of those drivers. Zynga's games *Farmville* and *Mafia Wars* are particularly successful examples of this kind of game. Using Facebook as a networking platform, these games are played in a single-player mode, but allow for cooperation and conflict in an asynchronous fashion. The player's performance is made public (within the player's Facebook Friends network) in some way, and this appears to be sufficient to keep people playing.

Whether this format is sufficient to motivate students to play the considerably more challenging games that would be required to develop mathematical thinking remains to be seen. *Farmville* itself is not really a game, and offers no mental challenges or any excitement at all as far as I can see. (I am definitely not one of the 75 million-plus individuals who apparently enjoy the game.) For educational purposes, such a game would not have to be a success in the open market. Actually, it's not a market, since social network games are virtually all made available on a play-for-free basis. The companies that make them earn revenue in different ways, such as offering a better version of the game at a price, selling virtual goods, selling mailing lists of players, etc. Students could be required to play a casual game by the math teacher. But if the game experience is not enjoyable—or satisfying—then all you have done is produce a more expensive, and glossier, version of old-fashioned math homework. The important goal of developing productive disposition in the students will have gone out the window.

Still, given the current market acceptance of social network games, the idea of creating and releasing a whole series of social network math ed video games is appealing. But don't expect it to happen overnight. Developing just one game activity that develops mathematical thinking skills is a long, hard process. Although one advantage of this approach is that a company can produce and release games serially, the real education value—and in particular interest from the education community—will come only when a sufficient critical mass of mathematical topics has been covered. Teachers, schools, education boards, and even states might be willing to pay for a suite of such games once there is sufficient curriculum coverage.

But releasing an online math ed video game of any kind is not the end of the process. In many ways, that is when things get really interesting for education researchers. This is due to a feature of online educational games that I have not addressed so far, but it is huge. When a player plays an online game, his or her every action can be recorded, including the time spent on each action or task. The data gathered can be used to tailor the challenges that the student/player will encounter next, and in some cases the in-game resources that will be made available to her. It can also be analyzed and tabulated and presented to the teacher, or—and this is the use I am going to end with—to the education researcher. Since the learning environment is digital, it is then possible for the researcher, working with the game developer, to modify the game to see how students respond to different kinds of curriculum or to test new educational hypotheses.

Online game developers frequently modify games in response to data collected from the logging of in-game player actions, from monitoring player fansites, and from direct feedback from players. Of course, they do it to eliminate bugs and to make the game more exciting to players so that they continue to play (and pay). The modifications are presented to the player for obligatory download at the next logon. In the case of an educational online game, in addition to making those kinds of modifications, we would also want to keep improving the education delivery. Besides making the game a better educational experience, researchers could also use it to examine features of situated learning and compare it with other forms of instruction.

An important thing to remember about education is that, insofar as it may be viewed as a science, it is the quintessential *empirical* science. Given all the theories about education that are thrown around, and all the pedagogic claims made, an outsider could be forgiven for thinking that anyone actually knows how people learn and what is the best way to teach them (or help them learn). We don't. Even when researchers set up a proper study to test a hypothesis, complete with a control group, it can be hard to come to any firm conclusion as to what was actually measured. For instance, much of the controversy that has surrounded Jo Boaler's research is due to this factor. Moreover, when a group of researchers works with a selected group of teachers to test some new classroom teaching method, you no longer have a typical classroom setting. The teachers of the new method will inevitably be interested in the new method, and excited to be part of the testing, and so will bring into a classroom a freshness and an enthusiasm that is not going to be present under more normal circumstances. Because they are doing something new, they will also put a fair amount of time and effort into becoming acquainted with the material, and be constantly monitoring the class for clues as to how it is going. The students also may become aware that they are part of an important study. Under those circumstances, it is almost inevitable that the results are going to be good. But is the new method contributing in any significant way to those good results, or are you just measuring what happens when a class is taught by well-informed, enthusiastic teachers who are sensitive to the class responses to the instruction? Though data collected from a video game is not going to eliminate those dangers of classroom experiments—in large part because even with the video game, classroom activity will still play a major role in the learning process—any source of new data, particularly impartial data (which game-collected data will be) can be useful.

Clearly, when it comes to using video games to revolutionize mathematics education, there is a lot that remains to be done. My purpose in writing this book is to try to persuade you, my fellow mathematicians and mathematics educators, to join in.

Suggested Further Reading

While articles in the popular press often claim (falsely) that video games kill traditional literacy, the same claim is patently false when applied to game designers and those who study video games. There is a wealth of information available on the design and use of video games: online articles and blogs, published research papers, and books (some of which are in the 600 to 1,000 pages range). As in any area, the quality varies. This short list is taken from the sources that I myself have found useful in learning about video games and their use in education, and in particular in writing this book.

Web Resources

There are a number of websites that provide excellent information about the design and use of video games. *Gamasutra* (http://www.gamasutra.com/) and *Terranova* (http://terranova.blogs.com/) are particularly good.

Anyone setting out to develop a mathematics education video game should definitely read Pascal Luban's article "Designing and Integrating Puzzles in Action-Adventure Games," which appeared on *Gamasutra* on December 6, 2002: http://www.gamasutra.com/features/20021206/luban_01.htm.

The New Media Consortium proves lots of information on the use of MMOs in education: http://www.nmc.org/horizonproject/2007/massively-multiplayer-educational-gaming.

Another excellent source for MMO design is http://update.multiverse.net/wiki/index.php/MMOG_Design.

Books

The one book I would recommend reading first for someone who wants to take a first look at the educational uses of video games is James Paul Gee's *What Video Games Have to Teach Us about Learning and Literacy* (Palgrave Macmillan, 2003). A more recent book by the same author is *Good Video Games and Good Learning (New Literacies and Digital Epistemologies)* (Peter Lang, 2007).

Anyone setting out to design an educational video game (or a video game in general) should definitely consult the two "bibles" of game design:

- Katie Salen and Eric Zimmerman, *Rules of Play: Game Design Fundamentals*, MIT Press, 2003 (688 pages);
- Katie Salen and Eric Zimmerman (eds.), *The Game Design Reader: A Rules of Play Anthology*, MIT Press, 2006 (923 pages).

Their focus is on games in general rather than just video games, and in an industry changing as rapidly as video games, much of what they contain now seems out of date. Moreover, some of it seems far removed from questions of implementation. But a careful read of both will yield a wealth of valuable insights.

Two other lengthy books that provide a lot of information are

- Richard Bartle, *Designing Virtual Worlds*, New Riders Games, 2003 (768 pages); and
- Jesse Schell, *The Art of Game Design: A Book of Lenses*, Morgan Kaufmann, 2008 (512 pages).

For general reading about video games in society and business, I would recommend the following three books:

- John Beck and Michael Wade, *Got Game, How the Gamer Generation is Reshaping Business Forever*, Harvard Business Press, 2004;
- Edward Castronova, *Synthetic Worlds: The Business and Culture of Online Games*, University of Chicago Press, 2005;
- Byron Reeves and J. Leighton Read, *Total Engagement: Using Games and Virtual Worlds to Change the Way People Work and Businesses Compete*, Harvard Business School Press, 2009.

# Bibliography

John C. Beck and Mitchell Wade, *Got Game: How the Gamer Generation Is Reshaping Business Forever*, Harvard Business Press, Boston, 2004.

Jo Boaler, *What's Math Got To Do With It?: How Parents and Teachers Can Help Children Learn to Love Their Least Favorite Subject*, Penguin Books, New York, 2009.

John Seely Brown, Allan Collins, and Paul Duguid, "Situated Cognition and the Culture of Learning," *Educational Researcher* 18 (1), 1989, pp. 32–41.

Noel Capon and Deanna Kuhn, "Logical Reasoning in the Supermarket: Adult Females' Use of a Proportional Reasoning Strategy in an Everyday Context," *Developmental Psychology*, 15 (4), July 1979, pp. 450–452.

James P Carse, *Finite and Infinite Games: A Vision of Life as Play and Possibility*, Ballantine Books, New York, 1987.

Edward Castronova, *Synthetic Worlds: The Business and Culture of Online Games*, University of Chicago Press, Chicago, 2006.

Terrence W. Deacon, *The Symbolic Species: The Co-evolution of Language and the Brain*, W. W. Norton, New York, 1997.

Stanislaus Dehaene, *The Number Sense: How the Mind Creates Mathematics*, Oxford University Press, Oxford, UK, 1997.

Keith Devlin, *The Language of Mathematics: Making the Invisible Visible*, W. H. Freeman, New York, 1998.

Keith Devlin, *The Math Gene: How Mathematical Thinking Evolved and Why Numbers Are Like Gossip*, Basic Books, New York, 2000.

Michele D. Dickey, "Game Design and Learning: A Conjectural Analysis of How Massively Multiplayer Online Role-Playing Games (MMORPGs) Foster Intrinsic Motivation," *Education Technology Research & Development*, 55 (3), 2007, pp. 253–273.

Thomas L. Friedman, *The World Is Flat: A Brief History of the Twenty-First Century*, Farrar, Straus and Giroux, New York, 2005.

James Paul Gee, *What Video Games Have to Teach Us About Learning and Literacy*, Palgrave Macmillan, New York, 2003.

Susan Gail Gerofsky, "The Word Problem as Genre in Mathematics Education," PhD diss., Simon Fraser University, Canada, 1999.

Malcolm Gladwell, *Blink: The Power of Thinking without Thinking*, Little, Brown and Company, Boston, 2005.

James Herndon, *How to Survive in Your Native Land*, Simon and Schuster, New York, 1971.

Jeremy Kilpatrick and Jane Swafford (eds.), *Adding It Up: Helping Children Learn Mathematics*, National Academies Press, Washington, DC, 2001.

George Lakoff and Raphael Núñez, *Where Mathematics Comes From: How the Embodied Mind Brings Mathematics into Being*, Basic Books, New York, 2000.

Jean Lave, *Cognition in Practice: Mind, Mathematics, and Culture in Everyday Life*, Cambridge University Press, Cambridge, UK, 1988.

Johnny Lott and Kathleen Nishimura (eds.), *Standards and Curriculum: A View from the Nation*, A Joint Report of the National Council of Teachers of Mathematics and the Association of State Supervisors of Mathematics, National Council of Teachers of Mathematics, Reston, VA, 2004.

John Mighton, *The Myth of Ability: Nurturing Mathematical Talent in Every Child*, Walker and Company, New York, 2004.

Terezinha Nunes and Peter Bryant, *Children Doing Mathematics (Understanding Children's Worlds)*, Wiley-Blackwell, New York, 1996.

Terezinha Nunes, Analucia Schliemann, and David Carraher, *Street Mathematics and School Mathematics*, Cambridge University Press, Cambridge, UK, 1993.

Byron Reeves and J. Leighton Read, *Total Engagement: Using Games and Virtual Worlds to Change the Way People Work and Businesses Compete*, Harvard Business School Press, Boston, 2009.

Laurence Sigler, *Fibonacci's Liber Abaci*, Springer-Verlag, New York, 2002.